Vibration-Based Techniques for Damage Detection and Localization in Engineering Structures

Computational and Experimental Methods in Structures

ISSN: 2044-9283

Series Editor: Ferri M. H. Aliabadi *(Imperial College London, UK)*

This series will include books on state-of-the-art developments in computational and experimental methods in structures, and as such it will comprise several volumes covering the latest developments. Each volume will consist of single-authored work or several chapters written by the leading researchers in the field. The aim will be to provide the fundamental concepts of experimental and computational methods as well as their relevance to real world problems.

The scope of the series covers the entire spectrum of structures in engineering. As such it will cover both classical topics in mechanics, as well as emerging scientific and engineering disciplines, such as: smart structures, nanoscience and nanotechnology; NEMS and MEMS; micro- and nano-device modelling; functional and smart material systems.

Published:

More information on this series can also be found at http://www.worldscientific.com/series/cems

(Continued at end of book)

Computational and Experimental Methods in Structures – Vol. 10

Vibration-Based Techniques for Damage Detection and Localization in Engineering Structures

Editors

Ali S Nobari

Imperial College London, UK
Amirkabir University of Technology, Iran

M H Ferri Aliabadi

Imperial College London, UK

World Scientific

NEW JERSEY · LONDON · SINGAPORE · BEIJING · SHANGHAI · HONG KONG · TAIPEI · CHENNAI · TOKYO

Published by

World Scientific Publishing Europe Ltd.

57 Shelton Street, Covent Garden, London WC2H 9HE

Head office: 5 Toh Tuck Link, Singapore 596224

USA office: 27 Warren Street, Suite 401-402, Hackensack, NJ 07601

Library of Congress Cataloging-in-Publication Data
Names: Nobari, Ali S., editor. | Aliabadi, M. H., editor.
Title: Vibration-based techniques for damage detection and localization in engineering structures /
    edited by Ali S Nobari (Imperial College London, UK & Amirkabir University of
    Technology, Iran), M.H. Ferri Aliabadi (Imperial College London, UK).
Description: [Hackensack] New Jersey : World Scientific, 2018. | Series: Computational and
    experimental methods in structures ; volume 10 | Includes bibliographical references and index.
Identifiers: LCCN 2017050793 | ISBN 9781786344960 (hc : alk. paper)
Subjects: LCSH: Structural health monitoring. | Vibration--Testing.
Classification: LCC TA656.6 .V53 2018 | DDC 624.1/71--dc23
LC record available at https://lccn.loc.gov/2017050793

British Library Cataloguing-in-Publication Data

A catalogue record for this book is available from the British Library.

For any available supplementary material, please visit
http://www.worldscientific.com/worldscibooks/10.1142/Q0145#t=suppl

Desk Editors: Anthony Alexander/Jennifer Brough/Koe Shi Ying

Typeset by Stallion Press
Email: enquiries@stallionpress.com

Preface

Structural health monitoring (SHM) and damage detection techniques are increasingly being taken up as a means of condition-based monitoring in offshore, civil, mechanical, and aeronautical engineering, both to improve safety and cost savings on maintenance. The need to be able to detect damage in complex structures subject to complex operational and environmental loads has led to the development of a wide range of techniques, of which many are based upon structural vibration analysis. This book brings together the latest trends and techniques related to the vibration-based damage detection. It provides a comprehensive coverage of the modern methodologies used for the vibration-based SHM. The main focus is on the techniques which are based on exploiting the medium-frequency range vibration response of structures.

Both time and frequency domain techniques are considered in the book. The first two chapters of the book deal with the data-based SHM method and the data processing techniques used for this purpose. Chapter 1 presents a detailed study of the various machine learning techniques applicable to the subject matter. The performance of these techniques are subsequently studied and compared in this chapter. As an extension to traditional principal component analysis (PCA), Chapter 2 describes the application of more robust singular spectrum analysis (SSA) technique for fault detection and examines its performance through case studies. Chapters 3 and 4 focus on the application of the global modal parameters for fault detection, with Chapter 3 mainly covering the composite structures and the application of modal damping as a fault detection tool.

Chapters 5–7 present SHM techniques based on the local modal parameters, and in particular, mode shape curvature analysis.

The editors wish to thank all the authors for their valuable contributions.

About the Editors

Ali Salehzadeh Nobari is a Visiting Professor at Aeronautics Department and Mechanical Engineering Department of Imperial College since 2010. He is also a Professor of Vibration Engineering at Aerospace Engineering Department of Amirkabir University of Technology (AUT) since 2008 and also the Head of Vibration Measurement, Signal Processing and Experimental Modal Analysis (EMA) Lab in AUT since 1999. He has more than 50 publications in ISI journals and 30 publications in international and national conferences. His area of research is vibration-based SHM, vibration measurement and signal processing, EMA, nonlinear dynamic system identification and material modeling, structural dynamic model updating, and identification of material models for viscoelastc materials (adhesives).

M. H. Ferri (Fraydon) Aliabadi is the Chair of Aerostructures and the Head of Department of Aeronautics, Imperial College London. He also holds the position of Zaharoff Professor in Aviation. He is a fellow of Royal Aeronautical Society. He has around 500 publications in the areas of computational structural mechanics, acoustics, fracture mechanics and fatigue, structural health monitoring, contact mechanics, optimisation, materials modeling, and boundary element methods. Professor Aliabadi is the Editor of *International Journal of Multiscale Modeling* and serves on editorial boards of several other journals. He is Editor-in-Chief of book series *Computer and Experimental Methods in Structures and Advances in Fracture*.

Contents

Chapter 1

Machine Learning Algorithms for Damage Detection

Eloi Figueiredo[*,‡] and Adam Santos[†,§]

Faculty of Engineering
Universidade Lusófona de Humanidades e Tecnologias
Campo Grande 376, 1749-024 Lisbon, Portugal
†Faculty of Computing and Electrical Engineering
Universidade Federal do Sul e Sudeste do Pará, F. 17, Q. 4
L. E., 68505-080 Marabá, Pará
‡eloi.figueiredo@ulusofona.pt
§adamdreyton@unifesspa.edu.br

This chapter poses the damage detection for civil, mechanical, and aerospace structures in the context of a pattern recognition paradigm for structural health monitoring (SHM), where machine learning algorithms are essential to learn the structural behavior from experience, following the same principle of the human brain. These algorithms are especially relevant in cases where the damage-sensitive features extracted from the structural responses are affected by changes caused by operational and environmental variability and changes caused by damage. State-of-the-art machine learning algorithms are presented based on Mahalanobis squared distance (MSD), Gaussian mixture models (GMMs), principal component analysis (PCA), kernel principal component analysis (KPCA), and autoassociative neural network (ANN); the bioinspired algorithms are highlighted as promising algorithms to overcome some of the limitations of the more traditional ones. All the algorithms present different working principles and seek to generalize the normal structural condition in order to detect deviations, from the baseline condition, associated with damage. The applicability of the chosen algorithms for damage detection will be tested on standard data sets from the Z-24 Bridge in Switzerland. These data sets are unique as they combine 1-year monitoring from the baseline condition, when influenced by extreme operational and environmental variability, with realistic damage scenarios.

Keywords: SHM; Data-based; Unsupervised machine learning; Statistical pattern recognition; Mahalanobis squared distance (MSD); Gaussian mixture models (GMMs); Principal component analysis (PCA); kernel principal

component analysis (KPCA); Autoassociative neural network (AANN); The
bioinspired algorithms.

1. Introduction

Engineering structures such as bridges, buildings, roads, railways, tunnels,
dams, offshore oil platforms, power generation systems, rotating machinery, and aircraft are present and play a crucial role in modern societies,
regardless of culture, geographical location, or economic development. The
safest, economical, and most durable structures are those that are well
managed and maintained. Health monitoring represents an important tool
in management activities as it permits one to identify early and progressive
structural damage.[1] The massive data obtained from monitoring must be
transformed to meaningful information to support the planning and designing of maintenance activities, increase the safety, verify hypotheses, reduce
uncertainty, and widen the knowledge and insight concerning the monitored
structure.

Structural health monitoring (SHM) is certainly one of the most powerful tools for infrastructure management. The process of implementing
an autonomous damage identification strategy for civil, mechanical, and
aerospace engineering infrastructure is traditionally referred to as SHM.[2]
The SHM process involves the observation of a structure over time using
periodically sampled response measurements from an array of sensors, the
extraction of damage-sensitive features from these measurements, and the
statistical analysis of these features to determine the actual structural condition. Even though the damage identification hierarchy is composed of
five levels (Section 2.7), this chapter poses the SHM process mostly in the
context of the first level — damage detection.

For long-term SHM, the output of the monitoring process is periodically
updated, providing information regarding the ability of the structure to perform its intended function in light of the inevitable aging and degradation
resulting from operational and environmental variability.[3] After extreme
events such as earthquakes or blast loadings, SHM is used for rapid condition screening and aims to provide, in nearly real time, reliable information
regarding the integrity of the structure.

There are arguably two main approaches to SHM: physics-based and
data-based. The physics-based approach uses the inverse problem technique
to calibrate numerical models (e.g. finite element models) and attempts
to identify damage by relating the measured data from the structures to

the estimated data from the models. On the other hand, the data-based approach is rooted in the machine learning field, where machine learning algorithms are essential to learn (or to model) the structural behavior from the experience (or past data), following the same principle of the human brain, and to perform pattern recognition for damage identification. They can be applied in supervised or unsupervised learning.[4] In the SHM field, supervised learning refers to the case where data from undamaged and damaged conditions are available to train the algorithms. Unsupervised learning refers to the case where training data are only available from the undamaged condition. Note that for high capital expenditure structures, such as most civil infrastructure, unsupervised learning algorithms are often required because only data from the undamaged condition are normally available.

Several unsupervised approaches have been proposed to detect structural damage by combining pattern recognition and machine learning.[5-7] This combination is often accomplished via a statistical pattern recognition (SPR) paradigm that establishes two phases:

(1) Learn a model which comprises undamaged data from the normal structural condition, considering nearly all operational and environmental influences.
(2) Test the learned model by classifying new undamaged or damaged data.

It is important to emphasize that, currently, most of the techniques used in the SPR are output-only, i.e. only the damage-sensitive features (often derived from vibration response measurements[8]) need to be measured, not the operational and environmental parameters.

In this chapter, several unsupervised machine learning algorithms for structural damage detection are described in the context of the SPR for SHM. State-of-the-art machine learning algorithms are presented based on Mahalanobis squared distance (MSD), Gaussian mixture models (GMMs), principal component analysis (PCA), kernel principal component analysis (KPCA), and autoassociative neural network (AANN); the bioinspired algorithms are highlighted as promising algorithms to overcome some of the limitations of the more traditional ones. These algorithms are relevant in cases where features extracted from the structural responses are affected by changes caused by operational and environmental variability and changes caused by damage. The applicability of the chosen algorithms for damage detection will be tested on standard data sets from the Z-24 Bridge in Switzerland. These data sets are unique as they combine 1-year monitoring from the baseline condition, when influenced by extreme operational and environmental variability, with realistic damage scenarios.

2. Statistical Pattern Recognition Paradigm for Structural Health Monitoring

2.1. *General definition*

It is believed that all approaches to SHM, as well as all traditional non-destructive techniques, can be posed in the context of a SPR paradigm.[2] Thus, the SPR paradigm for the development of SHM solutions is described as a four-phase process as illustrated in Figure 1.

The main objective of the SPR paradigm, in the context of SHM applications, is to recognize and distinguish between the patterns related to the undamaged structure condition under operational and environmental influences and those associated with the same structure under the damaged condition, by starting from sensor measurements of the monitored structure and finishing with the assessment of the actual structural condition. The four phases of the SHM process, as well as other background issues, are discussed in the following sections, presenting the main challenges associated with each phase.

2.2. *Operational evaluation*

The first phase for developing the SHM capability is to perform an operational evaluation. This part of the SHM process attempts to

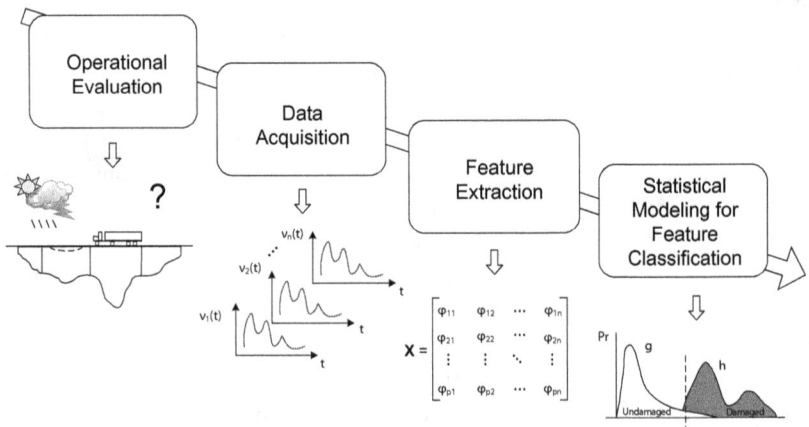

Figure 1.   The SHM process based on the SPR paradigm.

answer four essential questions regarding the implementation of an SHM system[9]:

- What are the life safety and/or economic justifications for monitoring the structure?
- How is damage defined for the structural system being monitored?
- What are the operational and environmental conditions under which the structural system of interest operates?
- What are the limitations on acquiring data in the operational environment?

The operational evaluation phase defines, and to the greatest extent possible quantifies, the damage that is to be identified. It also establishes the benefits to be gained from the deployment of the SHM system. This process also begins to impose limitations on what will be monitored and how to perform the monitoring, as well as adapting the monitoring to the unique aspects of the structural system and unique features of the damage that is to be identified and analyzed.

The main challenges of the operational evaluation phase are presented in the following:

- Most high-capital-expenditure civil engineering structures, such as bridges, are one-of-a-kind systems, influenced by the physical environment where they are built; therefore, it is more difficult to incorporate lessons learned from other nominally similar structural systems to define anticipated damage;
- Structural designs are often driven by low-probability, but extreme-impact events, such as earthquakes, hurricanes, terrorist actions, or floods;
- Generally, structural systems degrade slowly under normal use: corrosion and fatigue cracking, freeze–thaw/thermal damage, loss of pre-stressing forces, vibration-induced connectivity degradation, and hydrogen embrittlement;
- There is no widely accepted procedure yet to demonstrate the rate of return of the investment in an SHM system.

2.3. *Data acquisition*

The data acquisition phase of the SPR paradigm for SHM involves selecting the excitation methods; the sensor types, numbers, and locations; and the data acquisition/storage/processing/transmittal hardware. The interval at which the data should be collected (e.g. daily or hourly) is another consideration that must be addressed. The actual implementation of the data acquisition systems is application-specific, where economic issues will play a fundamental role in making the aforementioned choices.

A crucial premise regarding sensing and data acquisition is that these systems do not measure damage. Rather, they measure the response of a structure to its operational and environmental loading or the response to inputs from actuators embedded with the sensing system.[10] Depending on the sensing technology and the type of damage to be identified, the sensor readings may be more or less directly correlated to the presence and location of damage. Data interrogation procedures (feature extraction and statistical modeling for feature classification) are necessary components of an SHM system. They convert the sensor data into information about the structural condition. Moreover, to achieve successful SHM, the data acquisition systems have to be developed in conjunction with those data interrogation procedures.

The main challenges of the data acquisition phase are listed below:

- There is no sensor that measures damage. However, it is not possible to implement the SHM process without sensing;
- Definition of the data to be acquired and the data to be used in the feature extraction process: types of data to be acquired, sensor types and locations, bandwidth and sensitivity (e.g. dynamic range), data acquisition/transmittal/storage system, power requirements (e.g. energy harvesting), sampling frequencies, processor/memory requirements, excitation source (e.g. active sensing), sensor diagnostics, and so on;
- Number of sensors. Instrumenting large structures with lots of sensors still represents a sparsely instrumented system. Large sensor systems pose many challenges for reliability and data management;
- Ruggedness of sensors. Sensing systems must last for many years with minimal maintenance. The existence of harsh environments (e.g. thermal, mechanical, moisture, radiation, and corrosion) compromises the sensor durability. Need of sensor diagnostic capability;
- The sensing systems have evolved using ready-to-use off-the-shelf technology; however, the SHM research should pursue tailor-made SHM systems as a function of type of structure;
- The sensing system must be developed integrally with the feature selection and extraction as well as feature classification.

2.4. *Feature extraction*

A damage-sensitive feature is some quantity extracted from the structural response data which is correlated with the presence of damage in a structure (e.g. modal parameters,[11] quasi-static strains,[12] auto-regressive model parameters and residual errors,[5] local flexibilities,[13] and electromechanical impedances[14]). Ideally, a damage-sensitive feature will change in some consistent manner as the level of damage increases. Identifying features that can accurately distinguish a damaged structure from an undamaged one is the focus of most SHM technical literature. Fundamentally, the feature extraction process is based on fitting some model, either physics- or data-based, to the measured response data. The parameters of the models, or the predictive errors associated with them, become the damage-sensitive features. An alternative approach is to identify features that directly compare the sensor waveforms (e.g. influence lines and acceleration time series) or spectra of these waveforms (e.g. power spectra density) measured before and after damage. Many of the features identified for impedance-based and wave propagation-based SHM studies fall into this category.[15–18]

In the feature extraction phase, it is imperative to derive damage-sensitive features correlated with the severity of damage present in monitored structures in order to minimize false judgments in the following classification phase. Nevertheless, in real-world SHM applications, operational and environmental effects can mask damage-related changes in the features as well as alter the correlation between the magnitude of the features and the damage level. Commonly, the more sensitive a feature is to damage, the more sensitive it is to changing in the operational and environmental conditions (e.g. temperature and wind speed). To overcome this impact, the primary properties of damage-sensitive features are defined as follows:

- *Sensitivity* — A feature should be sensitive to damage and completely insensitive to everything else, which rarely occurs in practical SHM applications.
- *Dimensionality* — A feature vector should have the lowest dimension possible; high dimensionality induces undesirable complexity into the statistical models and storage mechanisms.
- *Computational requirements* — Minimal assumptions and minimal processor cycles, which facilitates the embedded systems.
- *Consistency* — Features magnitude should change monotonically with damage level.

One wants to use the simplest feature to distinguish between the damaged and undamaged structural system. However, there are a couple of challenges for feature selection and extraction, as described below:

- Feature selection is still based almost exclusively on engineering judgment;
- Quantifying the features' sensitivity to damage[19];
- Quantifying how the feature changes with the level of damage[19];
- Understanding how the feature varies with changing environmental and operational conditions.

2.5. *Statistical modeling for feature classification*

The development of statistical models to enhance the damage detection process is the phase concerned with the implementation of machine learning algorithms to normalize the data and to analyze the distributions of the extracted features in an effort to determine the structural condition.[20]

The machine learning algorithms used in statistical model development usually fall into three general categories: (i) group classification,[21,22] (ii) regression analysis,[23,24] and (iii) outlier detection.[5,7] The appropriate algorithm to use will depend on the ability to perform supervised or unsupervised learning. Herein, the unsupervised learning algorithms are considered because currently for high capital expenditure structures only data from the undamaged condition are available. Therefore, the algorithms fully described in Section 3 are used for outlier detection based on output residual errors or some sort of distance metric. In general, the machine learning algorithms presented in Section 3 usually output a feature vector of residuals with dimension equal to the dimension of the original feature vector. For instance, from Equations (12) and (24), a quantitative measure of damage for each feature vector can be established in the form of score. Thereby, a damage indicator (DI) might be adopted as the squared root of the sum-of-square errors (i.e. Euclidean norm) for each residual feature vector. Thus, a DI for each feature vector f ($f = 1, 2, \ldots, l$) of a test matrix $\mathbf{Z}$ is given by

$$\mathrm{DI}(\mathbf{z}_f) = \| \mathbf{e}_f \| \tag{1}$$

If the f feature vector is related to the undamaged condition, $\mathbf{z}_f \approx \hat{\mathbf{z}}_f$ and $\mathrm{DI}(\mathbf{z}_f) \approx 0$. On the other hand, if the feature vector comes from the damaged condition, the residual errors increase, and the DI deviates from zero, indicating an abnormal condition in the structure. A threshold can be adopted as an *ad hoc* distance or assuming a certain confidence interval.

In order to take into account variability and uncertainty and to classify those DIs that significantly deviates from zero, it is necessary to establish confidence intervals. When a considerable number of representative data is available from the undamaged condition, one might simply estimate thresholds based on values corresponding to a certain percentage of confidence over the training data. Therefore, multivariate outliers can then simply be defined as tests having DIs beyond a specific threshold.

Alternatively, for an outlier detection approach, a hypothesis test can be established, where the null hypothesis, H_0, is the undamaged condition and the alternative hypothesis, H_1, is presumably the damaged condition. To determine whether a feature vector is from a structure within the undamaged condition, a single-dimensional measure of separation between a new feature vector and an existing distribution is established. For instance, in Equation (4), if a multivariate feature vector $\mathbf{z}$ is extracted from the undamaged condition that corresponds to a multivariate Gaussian random distribution, then the DI will be Chi-square distributed with d degrees of freedom,

$$\mathrm{DI} \sim \chi_d^2 \tag{2}$$

When d increases, the probability density function (PDF) begins to approach a normal PDF, as predicted by the central limit theorem. Therefore, multivariate outliers can simply be defined as observations having DIs above a certain level. The assumption of a Chi-square distribution is indispensable for outlier detection because it permits one to define a cut-off value or threshold, c, for a level of significance, α, in the form of

$$c = \mathrm{inv}F_{\chi_n^2}(1-\alpha) \tag{3}$$

where $F_{\chi_d^2}$ is the cumulative distribution function of the central Chi-square distribution. Thus, a feature vector is considered to be a multivariate outlier (the null hypothesis is rejected) when its DI is equal to or greater than c. The selection of α carries a trade-off between the Type I error (false-positive indication of damage) and the Type II error (false-negative indication of damage). A level of significance equal to 5% is normally acceptable.

The performance assessment of damage detection is a fundamental aspect for evaluating and comparing models, algorithms, or classifiers. For the two-class problem in SHM (binary classification), in which the two sets of cases are labeled as damaged (or positive, P) or undamaged (or negative, N), assuming a given threshold, there are four possible outcomes as summarized in Figure 2 and Table 1. For a positive outcome, the case can

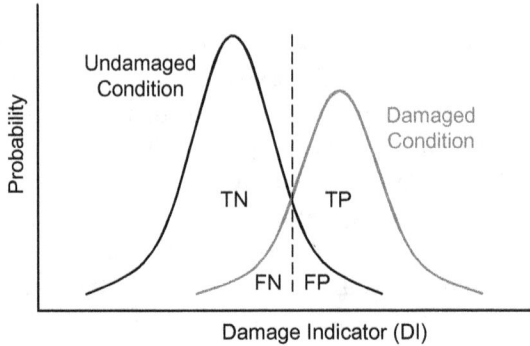

Figure 2.   Distributions from the undamaged and damaged conditions.

Table 1.   Accuracy of binary classification.

Outcome	Observed		
	Positive	Negative	Total
Positive	True positive (TP)	False positive (FP)	TP + FP
Negative	False negative (FN)	True negative (TN)	FN + TN
Total	TP + FN	FP + TN	TP + FN + FP + TN

be either true positive (TP) if the observed is positive or false positive (FP) if the observed is negative. On the other hand, for a negative outcome, the case can be either false negative (FN) if the observed is positive or true negative (TN) if the observed is negative. The misclassifications FP and FN are also known as Type I error and Type II error, respectively.

Therefore, false indications of damage fall into two categories: (i) false-positive damage indication (indication of damage when none is present, Type I error) and (ii) false-negative damage indication (no indication of damage when damage is present, Type II error). Errors of the first type are undesirable, as they will cause unnecessary downtime and consequent loss of revenue as well as loss of confidence in the monitoring system. More importantly, there are clearly safety issues if misclassifications of the second type occur.[1] Pattern recognition algorithms allow one to weigh one type of error above the other; this weighting may be one of the questions answered at the operational evaluation phase.

Additionally, receiver operating characteristic (ROC) curves provide a comprehensive and graphical manner to summarize the performance of classifiers.[25] The ROC curves were introduced in signal detection theory by electrical and radar engineers during World War II for detecting enemy objects in battle fields. Since that time, the ROC curves have become increasingly common in fields such as finance, atmosphere science, and medicine. In the field of machine learning, these curves have become a standard tool to evaluate the performance of binary classifiers.

Some of the challenges in the statistical modeling for feature classification are discussed in the following:

- The damage detection classification is currently posed in the context of FP and FN indications of damage. This technique recognizes that an FP classification may have different consequences than FN ones. Therefore, analytical approaches to defining threshold levels must: balance tradeoffs between FP and FN indications of damage, minimize FP when economic concerns drive the SHM applications, and minimize FNs when life-safety issues drive the SHM systems;
- Updating statistical models as new data become available;
- Managing the large volumes of data that will be produced by an online monitoring system;
- The choice of the machine learning algorithm for a specific application must be done as a function of the damage-sensitive feature used as well as the distribution of the features in the feature space.

2.6. *Data normalization, cleansing, fusion, and compression*

Inherent in the data acquisition, feature extraction, and statistical modeling phases of the SHM process are data normalization, cleansing, fusion, and compression.

Data normalization is the process that includes a wide range of steps for mitigating (or even removing) the effects of operational (e.g. traffic loading) and environmental (e.g. temperature) variations on the extracted features as well as for separating changes in damage-sensitive features caused by damage from those caused by varying operational and environmental conditions.[26] If not properly accounted for, operational and environmental factors can potentially result in false indications of damage. Therefore, this procedure usually contributes significantly to the structural damage detection process. Data normalization can be performed through the application of machine learning algorithms.

Data cleansing is the process of selectively choosing data to pass on to, or reject from, the feature selection process.[27]

Data fusion is the process of combining information from multiple sensors in an effort to enhance the reliability of the damage detection process and has been widely applied in SHM field.[28]

Data compression is the process of reducing the dimensionality of the data, or the features extracted from the data, in an effort to facilitate efficient storage of information and to enhance the statistical quantification of these parameters.[29]

The aforementioned four activities can be implemented in either hardware or software, and usually a combination of the two approaches is used.

2.7. *Hierarchical structure of damage identification*

The damage identification should be as detailed as possible to describe the damage impact on the structural system. In a broad sense, developments on damage identification can be broken down into three areas, namely damage detection, damage diagnosis, and damage prognosis. Nonetheless, damage diagnosis can be subdivided to better characterize the damage in terms of location, type, and severity. Thus, the hierarchical structure of damage identification can be decomposed in five levels (Figure 3) which answer the following questions.[9]

(1) Is the damage present in the system (detection)?
(2) Where is the damage (localization)?

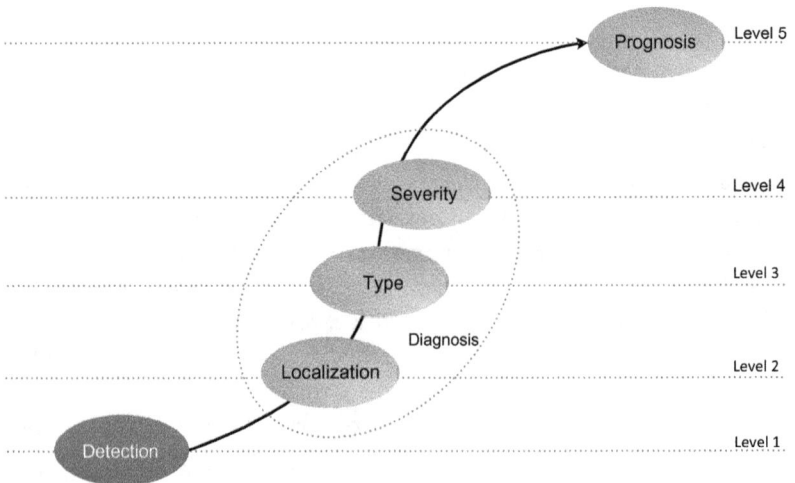

Figure 3.   Hierarchical structure of damage identification.

(3) What kind of damage is present (type)?

(4) What is the extent of damage (severity)?

(5) How much useful lifetime remains (prognosis)?

The answers to the questions above can only be made in a sequential manner. For example, the answer to the severity of damage can only be made with *a priori* knowledge of the type of damage. When applied in an unsupervised mode, machine learning algorithms are typically used to answer questions regarding the detection and localization of damage. When applied in a supervised learning mode and coupled with physics-based models, the statistical algorithms can be used to better determine the type of damage, the severity of damage, and the remaining useful lifetime. Note that damage prognosis at step five cannot be accomplished without an understanding of the damage accumulation process. For further discussion on the concept of damage prognosis, one should read the Reference.[30] Herein, only the first level and, for some extent, the fourth level are addressed.

3. Machine Learning Algorithms

In SHM, machine learning is the science of getting computers and algorithms to model the reality without knowing the physical laws of structures. In this section, several machine learning algorithms for data normalization and structural damage detection are described in the context of the SPR for SHM. These algorithms are especially relevant in cases where damage-sensitive features extracted from the structural responses are affected by changes caused by operational and environmental variability and changes caused by damage.

For general purposes, one should assume a training data matrix, $\mathbf{X} \in \mathbb{R}^{n \times d}$, with d-dimensional feature vectors from n different operational and environmental conditions when the structure is undamaged and a test data matrix, $\mathbf{Z} \in \mathbb{R}^{l \times d}$, where l is the number of feature vectors from the undamaged and/or damaged conditions.

3.1. *Mahalanobis squared distance*

The MSD is a distance measure for multivariate statistics outlier detection.[31] When one considers the training matrix, $\mathbf{X}$, with multivariate mean vector, $\boldsymbol{\mu}$, and covariance matrix, $\boldsymbol{\Sigma}$, the MSD (or DI in the context of SHM) between feature vectors from $\mathbf{X}$ and any new feature vector (or observation) from the test matrix, $\mathbf{Z}$, is computed as

$$\mathrm{DI}(\mathbf{z}) = (\mathbf{z} - \boldsymbol{\mu})\boldsymbol{\Sigma}^{-1}(\mathbf{z} - \boldsymbol{\mu})^{\top} \tag{4}$$

The assumption is that if a new observation is obtained from data collected on the damaged condition, which might include sources of operational and environmental variability, the observation is further from the mean of the normal condition. On the other hand, if an observation is obtained from a system within its undamaged condition, even with operational and environmental variability, this feature vector is closer to the mean of the normal condition. Under certain conditions, a threshold can be set up based on the assumption of Equation (2).

3.2. *Gaussian mixture model*

The GMM carries out a cluster-based model, using multivariate finite mixture models that aim to capture the main clusters/components of features, which correspond to the normal and stable state conditions of a bridge, even when it is affected by extreme operational and environmental conditions. Afterwards, an outlier detection strategy is implemented in relation to the chosen main components of states.[32] Basically, the damage detection is carried out based on multiples MSD-based algorithms, where the covariance matrices and mean vectors are functions of the main components.

Suppose that a training matrix $\mathbf{X}$ is available, arising from a mixture of Q distributions,[33]

$$f_{\mathrm{mix}}(\mathbf{x}) = \sum_{q=1}^{Q} \eta_q f_q(\mathbf{x} \,|\, \boldsymbol{\theta}_q) \tag{5}$$

with $f_q(\mathbf{x} \,|\, \boldsymbol{\theta}_q)$ being the density of a distribution from a known parametric distribution family $\tau(\boldsymbol{\theta})$. In this setting, one is concerned with the estimation of the component parameters $\boldsymbol{\theta} = (\boldsymbol{\theta}_1, \ldots, \boldsymbol{\theta}_q)$ and the mixture weights $\eta = (\eta_1, \ldots, \eta_q)$ of the underlying mixture distributions, based on the data $\mathbf{X}$. In this case, one assumes underlying multivariate Gaussian mixture distributions, and therefore each component density is a d-variate Gaussian function in form of[33]

$$f(\mathbf{x} \,|\, \boldsymbol{\mu}_q, \boldsymbol{\Sigma}_q) = \frac{\exp\{-\frac{1}{2}(\mathbf{x} - \boldsymbol{\mu}_q)^{\top}\boldsymbol{\Sigma}_q^{-1}(\mathbf{x} - \boldsymbol{\mu}_q)\}}{(2\pi)^{d/2}\sqrt{\det(\boldsymbol{\Sigma}_q)}} \tag{6}$$

with unknown parameters $\boldsymbol{\theta}_q = \{\boldsymbol{\mu}_q, \boldsymbol{\Sigma}_q\}$, namely mean vectors $\boldsymbol{\mu}_q$ and covariance matrices $\boldsymbol{\Sigma}_q$. The mixture weights are constrained to be $\sum_{q=1}^{Q} \eta_q = 1$. The complete GMM is parameterized by the mean vectors, covariance matrices, and mixture weights from all component densities, $\{\boldsymbol{\mu}_q, \boldsymbol{\Sigma}_q, \eta_q\}_{q=1,...,Q}$.

The parameters are estimated from the training data using the classical maximum likelihood (ML) estimation based on the expectation-maximization (EM) algorithm.[34] Other methods can be used to estimate the parameters of a GMM, as proposed in Ref. 35. The Bayesian information criterion (BIC)[36] is used to determine the appropriate number of components, which introduces a penalty term for the number of parameters in the model.

For the damage detection strategy, and for each observation $\mathbf{z}$, one needs to estimate Q DIs. Basically, for each main component q of the data,

$$\text{DI}_q(\mathbf{z}) = (\mathbf{z} - \boldsymbol{\mu}_q)\boldsymbol{\Sigma}_q^{-1}(\mathbf{z} - \boldsymbol{\mu}_q)^\top \tag{7}$$

where $\boldsymbol{\mu}_q$ and $\boldsymbol{\Sigma}_q$ represent all the observations from the q component of the data, when the structure is undamaged even though under varying operational and environmental conditions. Finally, for each observation, the DI is given by the smallest DI estimated on each component,

$$\text{DI}(\mathbf{z}) = \min[\text{DI}_q(\mathbf{z})] \tag{8}$$

3.3. *Principal component analysis*

PCA is a linear method for mapping multi-dimensional data (input space) into a lower dimension (feature space) with minimal loss of information.[37] Herein, PCA is used as a data normalization method,[38] assuming the training data matrix $\mathbf{X}$ decomposed in the form of

$$\mathbf{X} = \mathbf{T}\mathbf{U}^\top = \sum_{i=1}^{d} \mathbf{t}_i \mathbf{u}_i^\top \tag{9}$$

where $\mathbf{T}$ is called the scores matrix and $\mathbf{U}$ is a set of d orthogonal vectors, $\mathbf{u}_i$, also called the loadings matrix. The orthogonal vectors can be obtained by decomposing the covariance matrix of $\mathbf{X}$ in the form of $\boldsymbol{\Sigma} = \mathbf{U}\boldsymbol{\Lambda}\mathbf{U}^\top$, where $\boldsymbol{\Lambda}$ is a diagonal matrix containing the ranked eigenvalues $\lambda_{i,i}$, and $\mathbf{U}$ is the matrix containing the corresponding eigenvectors. The eigenvectors associated to the higher eigenvalues are the principal components (PCs) of the data matrix and they correspond to the dimensions that have the largest variability in the data. Basically, this method permits one to perform an orthogonal transformation or mapping, retaining only the PCs r ($\leq d$), also know as the number of factors. Precisely, choosing only the first r eigenvectors, the final matrix can be rewritten without significant loss of information in the form of

$$\mathbf{X} = \mathbf{T}_r\mathbf{U}_r^\top + \mathbf{E} = \sum_{i=1}^{r} \mathbf{t}_i\mathbf{u}_i^\top + \mathbf{E} \tag{10}$$

where $\mathbf{E}$ is the residual matrix resulting in the r factors. The coefficients of the linear transformation are such that if the feature transformation is applied to the data set and then reversed, there will be a negligible difference between the original and reconstructed data. The r factors may be automatically obtained by test the minimal percentage of the variance γ (usually from 0.9 to 0.95) to explain the variability in the matrix $\mathbf{X}$

$$\gamma \leq \frac{\sum_{i=1}^{r} \lambda_{i,i}}{\sum_{i=1}^{d} \lambda_{i,i}} \tag{11}$$

In the context of data normalization, the PCA algorithm can be summarized as follows: the loadings matrix is obtained from $\mathbf{X}$, the test matrix $\mathbf{Z}$ is mapped onto the feature space $\mathbb{R}^r$ and reversed back to the original space $\mathbb{R}^d$, the residual matrix $\mathbf{E}$ is computed as the difference between the original and the reconstructed test matrix

$$\mathbf{E} = \mathbf{Z} - (\mathbf{Z}\mathbf{U}_r)\mathbf{U}_r^\top \tag{12}$$

and finally to establish a quantitative measure of damage, the DIs can be calculated using, for example, the Euclidean distance on the residuals $\mathbf{E}$.

3.4. *Kernel principal component analysis*

The KPCA algorithm is the nonlinear improvement of the PCA.[39] Let $\mathcal{X} \in \mathbb{R}^d$ be the input space such that the observations $\mathbf{x}_i \in \mathcal{X}$, $i = 1,\ldots,n$. Every observation $\mathbf{x}$ is then mapped to a d_ϕ-dimensional feature space $\mathcal{H}$ by applying the mapping functions ϕ_m, $m = 1,\ldots,d_\phi$, where

$$\phi(\mathbf{x}) = [\phi_1(\mathbf{x}) \quad \phi_2(\mathbf{x}) \quad \cdots \quad \phi_{d_\phi}(\mathbf{x})]^\top \tag{13}$$

By employing the kernel trick,[40] $\mathcal{K} : \mathcal{X} \times \mathcal{X} \mapsto \mathbb{R}$ is defined as a positive semi-definite scalar kernel function satisfying for all $\mathbf{x}_i, \mathbf{x}_j \in \mathcal{X}$

$$\mathcal{K}(\mathbf{x}_i, \mathbf{x}_j) = \phi(\mathbf{x}_i)^\top \phi(\mathbf{x}_j) \tag{14}$$

$\mathcal{K}(\cdot)$ defines an inner product that allows one to map the observations implicitly to a high-dimensional kernel space. Let

$$\mathbf{\Phi} = [\phi(\mathbf{x}_1) \quad \phi(\mathbf{x}_2) \quad \dots \quad \phi(\mathbf{x}_n)] \tag{15}$$

be the $d_\phi \times n$ matrix of the mapped observations and $\mathbf{K} = \mathbf{\Phi}^\top \mathbf{\Phi}$ be the $n \times n$ kernel (Gram) matrix. According to Mercer's theorem, any continuous, symmetric, and positive semi-definite function that maps $(\mathbf{x}_i, \mathbf{x}_j)$ onto a high-dimensional feature space can represent a kernel.[41] The Kernel trick then consists of specifying the kernel $\mathcal{K}(\cdot)$ instead of the mapping ϕ. Herein, a Gaussian kernel is applied[42]

$$\mathcal{K}(\mathbf{x}_i, \mathbf{x}_j) = \exp\left(-\frac{\|\mathbf{x}_i - \mathbf{x}_j\|^2}{2\sigma^2}\right) \tag{16}$$

where the chosen nonlinear kernel implicitly defines a high-dimensional feature space with a bandwidth parameter σ^2.

To avoid the first eigenvector (or PC) becomes much larger than the other components, the kernel matrix $\mathbf{K}$ should be replaced by a centered version,[39]

$$\mathbf{K} \to \mathbf{K} - \frac{\mathbf{1_n}}{n}\mathbf{K} - \mathbf{K}\frac{\mathbf{1_n}}{n} + \frac{\mathbf{1_n}}{n}\mathbf{K}\frac{\mathbf{1_n}}{n} \tag{17}$$

with $\mathbf{1_n}$ as the $n \times n$ matrix where all elements are equal to 1.

The eigenvalues $\mathbf{\Sigma}$ and the corresponding eigenvectors $\mathbf{U}$ can be derived by using singular value decomposition (SVD) to solve the generalized eigenvalue problem,[39]

$$\mathbf{K}\mathbf{U} = \mathbf{U}\mathbf{\Sigma} \tag{18}$$

Afterwards, the $\mathbf{\Sigma}_1$ and $\mathbf{U}_1$ should be defined as follows:

$$\mathbf{\Sigma} = [\mathbf{\Sigma}_1 \quad \mathbf{\Sigma}_2], \ \mathbf{\Sigma}_1 \in \mathbb{R}^{r \times r} \tag{19}$$

$$\mathbf{U} = [\mathbf{U}_1 \quad \mathbf{U}_2], \ \mathbf{U}_1 \in \mathbb{R}^{n \times r} \tag{20}$$

where $\mathbf{\Sigma}_1$ comprises the r largest eigenvalues and $\mathbf{U}_1$ the corresponding eigenvectors.

There are several methods to optimize the bandwidth parameter σ^2 of the Gaussian kernel,[43] requiring only that $n \geq d$. The maximization of information entropy from kernel matrix, $\mathbf{K}$, is the most indicated method in the context of damage detection for SHM.

Also, several criteria have been proposed to determine the number of PCs r retained in the high-dimensional feature space.[37,44] For instance, r can be derived to comprise nearly all normal variability of the training data by using $\gamma = 0.99$ in Equation (11).

Note that the variance retained in the standard PCA is usually 0.9–0.95.[37] On the other hand, the KPCA has been often used with 0.99 because of the high-dimensional mapped feature space obtained with the Gaussian kernel, so there are potentially n non-zero PCs.

Considering that an undamaged data model was established in the training phase, in the test phase the DI is generated to any new observation $\mathbf{z}_i \in \mathbb{R}^d$, $i = 1, \ldots, l$. First, the new observation should be mapped onto the high-dimensional feature space in the form of $\mathbf{\Phi}(\mathbf{z}_i)^\top \mathbf{\Phi}$ (or $\mathbf{\Phi}^\top \mathbf{\Phi}(\mathbf{z}_i)$), by using $\mathbf{X}$ and $\mathbf{z}_i$ in Equation (16). Besides, a centering should be performed, for instance, such as,

$$\mathbf{\Phi}(\mathbf{z}_i)^\top \mathbf{\Phi} \to \mathbf{\Phi}(\mathbf{z}_i)^\top \mathbf{\Phi} - \frac{\mathbf{\check{I}_n}}{n}\mathbf{K} - \mathbf{\Phi}(\mathbf{z}_i)^\top \mathbf{\Phi}\frac{\mathbf{1_n}}{n} + \frac{\mathbf{\check{I}_n}}{n}\mathbf{K}\frac{\mathbf{1_n}}{n} \qquad (21)$$

with $\mathbf{\check{I}_n}$ as the $l \times n$ matrix where all elements are equal to 1.

Second, the eigenvectors $\mathbf{U}_1$ should be replaced by a normalized version,

$$\mathbf{u}_m \to \frac{\mathbf{u}_m}{\sqrt{\mathbf{\Sigma}_{m,m}}}, \quad m = 1, \ldots, r \qquad (22)$$

Finally, a DI is generated for the lth new observation as follows:

$$\mathrm{DI}(\mathbf{z}_l) = \mathbf{\Phi}(\mathbf{z}_l)^\top \mathbf{\Phi}\mathbf{U}_1\mathbf{U}_1^\top \mathbf{\Phi}^\top \mathbf{\Phi}(\mathbf{z}_l) \qquad (23)$$

3.5. *Autoassociative neural network*

The AANN-based algorithm, another type of nonlinear PCA, is trained to characterize the underlying dependency of the identified features on the unobserved operational and environmental factors by treating this unobserved dependency as hidden intrinsic variables in the network architecture.[45,46] The AANN architecture consists of three hidden layers: the mapping layer, the bottleneck layer, and de-mapping layer. More details on the network, including the number of nodes to use, can be found in Ref. 47.

In the context of data normalization for SHM, the AANN is first trained to learn the correlations between features from the training matrix, $\mathbf{X}$. Then the network should be able to quantify the unmeasured sources of variability that influence the structural response. This variability is represented at the bottleneck output, where the number of nodes (or factors) should correspond to the number of unobserved independent factors that influence the structural response. Second, for the test matrix, $\mathbf{Z}$, the residual matrix $\mathbf{E}$ is given by

$$\mathbf{E} = \mathbf{Z} - \tilde{\mathbf{Z}} \tag{24}$$

where $\tilde{\mathbf{Z}}$ corresponds to the estimated feature vectors that are the output of the network. The DIs can be calculated using, once again, the Euclidean distance on the residuals $\mathbf{E}$.

Notice that this algorithm is a mixture of two different learning approaches, i.e. supervised learning is used to obtain the operational and environmental conditions dependency albeit without direct measurement of these conditions, while unsupervised learning is used to detect damage.[5] It should be re-emphasized that a key issue is to appropriately define the number of nodes in the bottleneck layer, which depends on the independent sources of variability present in the measurements and influences the damage detection performance.

3.6. *Bioinspired algorithms*

More recently, a new class of cluster-based methods inspired on genetic algorithms (GAs) and particle swarm optimization (PSO)[48–50] has been developed to model the normal condition of structures as data clusters and then estimate the actual structural condition based on the learned clusters, similar to the GMM-based approach described above. The bioinspired algorithms attempt to overcome the limitation of the GMM algorithm related to premature convergence toward a local optimal, which is a result of its dependence on the choice of initial parameters.[51]

3.6.1. *Memetic algorithms*

Two bioinspired versions of the GMM-approach can be derived by combining the GA and PSO with the EM algorithm in the form of a memetic algorithm (MA) or global EM (GEM-GA or GEM-PSO).[48,49] First, the GEM approach applies the BIC as a fitness function to improve the performance of the EM algorithm via an MA, which consequently provides a robust generalization of the normal condition of a structure. Second, damage detection strategies based on the Mahalanobis or Euclidean distances can be applied. MAs are population-based metaheuristics composed of an evolutionary framework and a set of local search algorithms.[52] A general MA can be defined as follows[53]:

(1) *Initialize* a population of candidate solutions P_1.
(2) While a *termination criterion* is not satisfied, repeat.

    (i) *Cooperate* between candidate solutions from P_1 to generate a new population P_2.
    (ii) *Improve* the candidate solutions from P_2 to generate a new population P_3.
    (iii) *Compete* within the set $P_1 \cup P_3$ to generate a new population P_1 for the next generation.
    (iv) If P_1 converges, *Restart* some chosen solutions.

(3) Return the *best solution* encountered.

The *Initialize* procedure produces the initial set of random candidate solutions as high-quality solutions generated by applying a local search algorithm, and the *termination criterion* usually verifies the total number of generations and/or a maximum number of generations without improvement. The *Cooperate* procedure arises on the selection and combination, determining the solutions that will be merged to create new promising solutions. The *Improve* procedure applies a local search method on new solutions derived in the *Cooperate* procedure. The *Compete* procedure updates the current population using the old population and the new population, determining the solutions that will survive in the following generations. To overcome premature convergence to suboptimal regions of the search space, the *Restart* procedure acts as a corrective measure on the population.[53]

Hence, MAs comprise notions from population-based global search and local search methods. In this study, the MA is a hybrid algorithm that combines GA and PSO (global) with EM algorithm (local). Hereafter, for instance, the general framework of the GEM–GA approach is presented, and its parameters and operators are discussed in more detail:

(1) *Initialize* $P_1(a)$, $a = 0$, $b = 0$, $o = 0$.
(2) $(P_2(a), \mathrm{BIC}_1) \leftarrow$ perform R EM steps on $P_1(a)$.
(3) While $(a \leq \hat{a})$ and $(b \leq \hat{b})$ is satisfied, repeat.

(i) *Cooperate* such as $(P_3(a))$ $\leftarrow$ recombine on $P_2(a)$ and after $(P_4(a))$ $\leftarrow$ mutate on $P_3(a)$.

(ii) *Improve* $P_4(a)$ such as $(P_5(a), \text{BIC}_2)$ $\leftarrow$ perform R EM steps on $P_4(a)$.

(iii) *Compete* between the populations $P_2(a)$ and $P_5(a)$ such as $(P_2(a+1), \text{BIC}_1, b_{\text{GMM}}, \text{BIC}_{\min})$ $\leftarrow$ select on $P_2(a)$ and $P_5(a)$ via BIC_1 and BIC_2.

(iv) Increase $a = a + 1$, and $o = o + 1$ if and only if there is no improvement in $\text{BIC}_{\min}$.

(v) If $o = 100$, then *Restart* the worst 90% of individuals from $P_2(a)$, increase $b = b + 1$ and set $o = 0$.

(4) *Improve* b_{GMM} if and only if it is not converged, such as $(b_{\text{GMM}}, \text{BIC}_{\min})$ $\leftarrow$ perform EM steps on b_{GMM} until convergence of the LogL is reached.

(5) Return b_{GMM} as the *best solution* encountered.

In the GEM–GA approach, each individual in the population represents a candidate solution of the GMM, i.e. a set of parameters, Θ, that specifies a GMM. Thus, an individual is composed of two different parts. The first part indicates whether a component is activated ($[0.5, 1]$) or not ($[0, 0.5]$) for learning the GMM, and the length of this part is the maximum number of activated components $Q_{\max}$. The parameters, the mean vector μ_q and covariance matrix Σ_q, of $Q_{\max}$ components, are represented in the second part. Each component includes $(d^2 + 3d)/2$ parameters. Note that the covariance matrix, Σ_q, must be symmetric, thus only the upper (or lower) triangular part of the matrix is encoded in the individual.

First, a random population P_1, the number of generations a, restarts b, and generations without improvement o are initialized. Afterwards, by applying R EM steps on P_1, with $R = 20$, initial high-quality solutions and their fitness values are derived as P_2 and BIC_1, respectively. The evolutionary process of the GEM–GA is stopped when $a = \hat{a}$ or $b = \hat{b}$, then the convergence of the best solution, b_{GMM}, found so far is checked. If it converged in terms of LogL, the final solution is the current best solution, otherwise the EM algorithm improves b_{GMM} until convergence is reached.

The *Cooperate* procedure includes the parent selection, recombination and mutation. For parent selection, the well-know binary tournament is used on the population P_2 to select parents for recombination.[54] In turn, the recombination merges the chosen parents to generate offspring individuals with the crossover probability $p_{cro} = 0.8$. The two-point crossover is applied, in which two crossover positions between $\{1, \ldots, Q_{max} - 1\}$ are randomly selected within first part of the parent individual. The values of the genes to the right of these positions are exchanged between parent individuals for the first part and this exchange is generalized to their associated parameters in the second part. A Gaussian mutation, with the mutation probability $p_{mut} = 0.05$ for each gene of each individual (excluding covariance matrices), is performed on the offspring population P_3 to yield the population P_4.

An improvement on the population P_4 is achieved by applying the EM algorithm that delivers the population P_5 and its fitness values BIC_2, which are used together with P_2 and BIC_1 to select the new population of survivors for the next generation as well as the best solution, b_{GMM}, and its fitness value BIC_{min}. This survival selection is based on the elitist $(\delta + \lambda)$-strategy.[55] If there is no improvement in BIC_{min} during 100 consecutive generations, the worst 90% of individuals from P_2 are restarted as high-quality solutions (generated by applying the EM algorithm) to explore other regions of the search space. More details related to the working principles and implementation of those bioinspired methods can be found in Figure 4 and Ref. 49, respectively.

3.6.2. *Genetic algorithm for decision boundary analysis*

Another bioinspired technique is a nonparametric approach based on a GA for decision boundary analysis (GADBA).[50] In this case, no underlying distribution is assumed and the genetic-based clustering approach is supported by a novel concentric hypersphere (CH) algorithm to regularize the number of clusters and mitigate the cluster redundancy, as illustrated in Figure 5; in this case, the CH algorithm is applied to a three-component scenario with a five-centroid candidate solution. Initially, in (a) the centroids are moved to better fit the data. In (b) and (c) two centroids are agglutinated. On the other hand, in (d) when the CH algorithm is stopped, only one cluster is defined. If the agglutination process occurs the remaining centroids are evaluated to avoid bad positioning.

The GADBA-based approach can be summarized in two steps[50]: (i) the main normal state conditions of a structural system are automatically discovered by clustering the training observations according to the closest centroids, which are targets of the optimization performed by the GA; this optimization defines boundary regions between the clusters and reduces

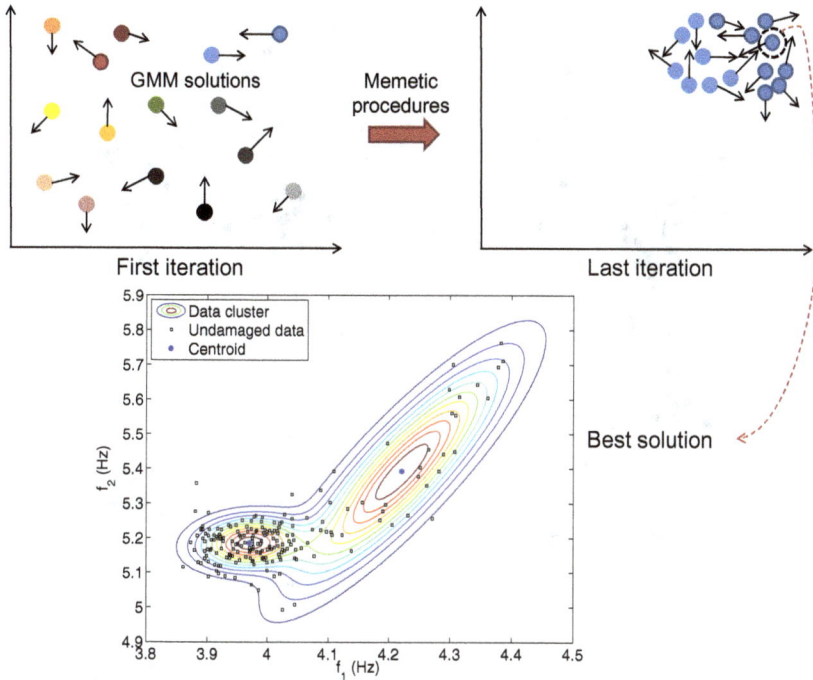

Figure 4. General overview of the GEM-GA (or GEM-PSO) approach. After a given number of iterations, by applying the memetic procedures, several GMM solutions with distinct number of data clusters converge to the same region in the search space where these initial candidate solutions become optimal solutions with approximately the same number of data clusters modeling the monitoring data. Afterwards, the best solution is selected to model the undamaged condition of a structure based on the BIC (in this case the solution is composed of two clusters).

the number of discovered state conditions; and (ii) the damage detection strategy is based on the Euclidean distance between a test observation and the optimized centroids. For each observation, the minimum distance to the centroids represents the DI.

4. Applicability on a Real-World Structure

In this section, the described SHM process, based on the SPR paradigm, is applied on real-world data sets acquired by monitoring a full-scale civil structure for almost a year, which combines the undamaged condition under normal variability with realistic damage scenarios.

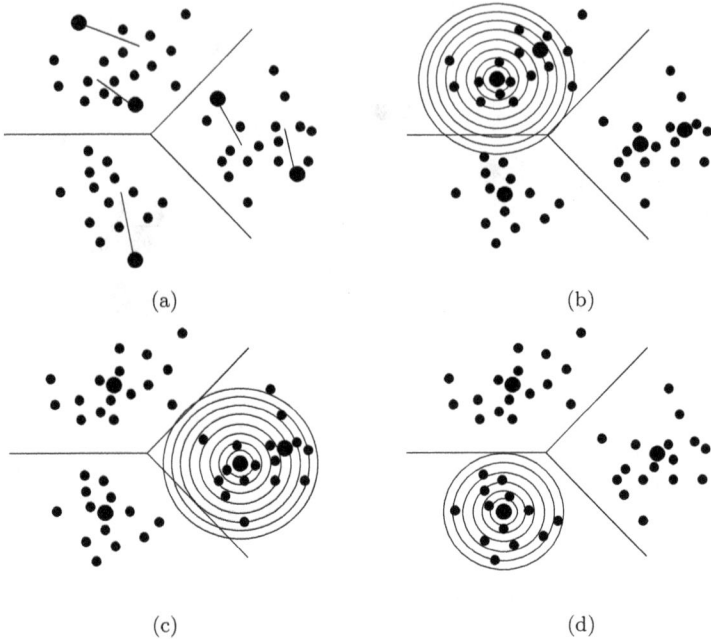

(a)   (b)

(c)   (d)

Figure 5. CH algorithm using linear inflation. (a) The centroids are moved to better fit the data, (b) and (c) two centroids are agglutinated, (d) when the CH algorithm is stopped, only one cluster is defined.[50]

4.1. Z-24 Bridge: Structural description, feature extraction and data sets

The Z-24 Bridge was a post-tensioned concrete box girder bridge composed of a main span of 30 m and two side-spans of 14 m, as shown in Figure 6. The bridge, before complete demolition, was extensively instrumented and tested with the purpose of providing a feasibility tool for vibration-based SHM in civil engineering.[56] A long-term monitoring program was carried out, from 11 November 1997 until 10 September 1998, to quantify the operational and environmental variability present on the bridge and to detect damage artificially introduced, in a controlled manner, in the last month of operation. Every hour, for 11 min, eight accelerometers captured the vibrations of the bridge and an array of sensors measured environmental parameters, such as temperature at several locations.

Progressive damage tests were performed in, roughly, 1-month time period before the demolition of the bridge to prove that realistic damage has

(a)

(b)

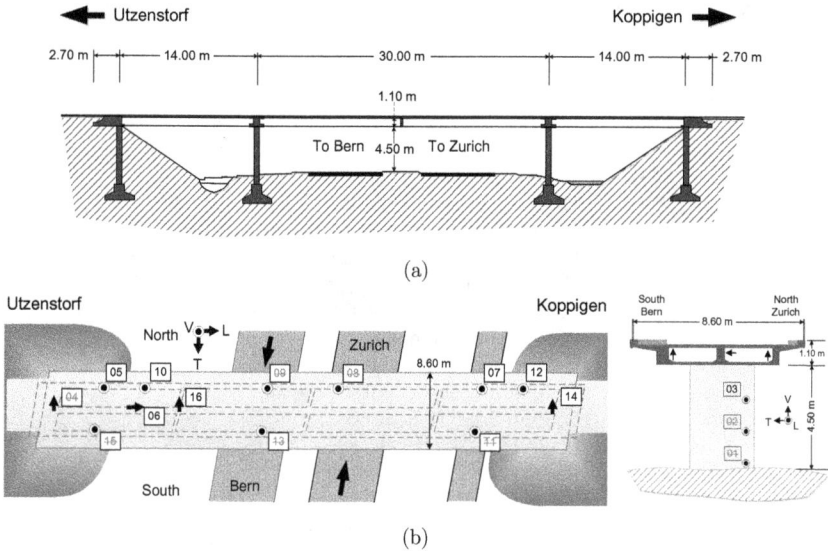

Figure 6. Longitudinal section (a) and the location and orientation of accelerometers (b) on the Z-24 Bridge.[58] Marked sensors failed during the monitoring campaign.

a measurable influence on the bridge dynamics,[59] as summarized in Table 2. Note that the continuous monitoring system was still running during the progressive damage tests, which permits one to validate the SHM system to detect accumulative damage on long-term monitoring.

In this case, the natural frequencies of the Z-24 Bridge are used as damage-sensitive features, as they are: extensively used in civil engineering structures, relatively straightforward to compute, and potentially consistent with damage level in a low-dimensional space. They were estimated using a reference-based stochastic subspace identification method on time series from the accelerometers.[57] The first four natural frequencies estimated, hourly, from 11 November 1997 to 10 September 1998, with a total of 3932 observations, are plotted in Figure 7. The first 3470 observations (11 November 1997 to 4 August 1998) correspond to the damage-sensitive feature vectors extracted within the undamaged condition (baseline condition or normal condition) under operational and environmental influences. The last 462 observations (5 August to 10 September 1998) correspond to the damage progressive testing period, which is highlighted, especially in the second frequency, by a clear drop in the magnitude of the frequency.

Table 2.   Progressive damage test scenarios.

Date[a]	Scenario description[b]
4 Aug 1998	Undamaged condition
9 Aug 1998	Installation of the PSS
10 Aug 1998	Lowering of pier, 2 cm
12 Aug 1998	Lowering of pier, 4 cm
17 Aug 1998	Lowering of pier, 8 cm
18 Aug 1998	Lowering of pier, 9.5 cm
19 Aug 1998	Lifting of pier, tilt of foundation
20 Aug 1998	New reference condition (after removal of the PSS)
25 Aug 1998	Spalling of concrete at soffit ($12\,m^2$)
26 Aug 1998	Spalling of concrete at soffit ($24\,m^2$)
27 Aug 1998	Landslide of 1 m at abutment
31 Aug 1998	Failure of concrete hinge
2 Sept 1998	Failure of 2 anchor heads
3 Sept 1998	Failure of 4 anchor heads
7 Sept 1998	Rupture of 2 out of 16 tendons
8 Sept 1998	Rupture of 4 out of 16 tendons
9 Sept 1998	Rupture of 6 out of 16 tendons

Notes:

[a]The dates refer to the additional vibration measurements.

[b]Consider pier settlement system as PSS.

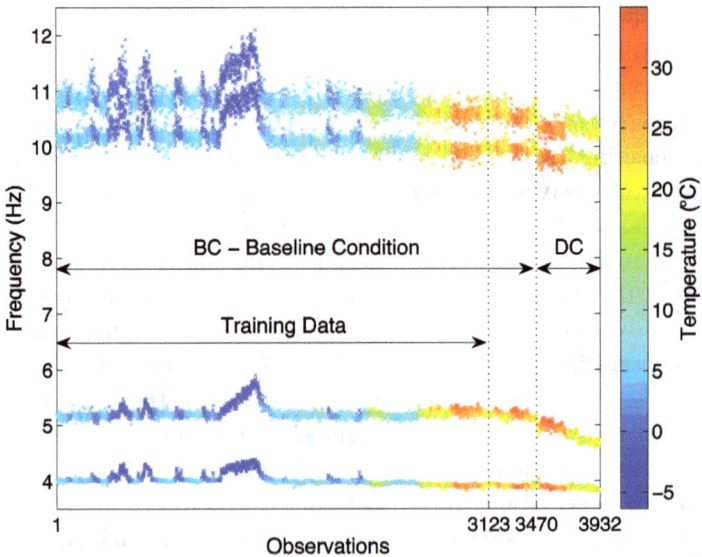

Figure 7.   First four natural frequencies of the Z-24 Bridge: 1–3470 baseline/undamaged condition (BC), 3471–3932 damaged condition (DC).

Note that the damage scenarios are carried out in a sequential manner, which cause a cumulative degradation of the bridge. The dimension reduction carried out through the feature extraction process permitted one to reduce the 12 GB of raw data to 100 kB, roughly, of structural information. For the baseline condition period, the observed jumps in the natural frequencies are associated to the asphalt layer, in cold periods, which contributes, significantly, to the stiffness of the bridge. This high influence of the temperature on the natural frequencies, as well as on the structure dynamics, may be considered a nonlinearity as the stiffness changes for temperatures below and above 0°C.[59]

The data sets from the Z-24 Bridge are particularly unique to test the applicability of vibration-based SHM methods because they encompass a wide spectrum of challenges encountered in practical SHM problems; for instance, significant structural changes caused by environmental factors (mainly temperature) and by several damage scenarios. During the undamaged condition of the bridge, high influence of the temperature on the natural frequencies is evident, as seen in Figure 7, and this influence is relatively larger than those caused by damage. This fact highlights the need for application of data normalization algorithms in order to attenuate the environmental variability before the damage detection step takes place.

For the sake of visualization, Figure 8 shows the first two natural frequencies in the 2D feature space. This dimension reduction still can be representative of the real structural state condition, as some of the features are highly correlated (for example, the first and the third natural frequencies are correlated with a correlation coefficient of 0.94). The figure shows that the structural condition of the bridge populates different regions in the feature space as a function of the ambient temperature and the presence of damage. Figure 9 shows the first two natural frequencies as a function of the ambient temperature, which highlights that linear approaches are generally liable to relatively poor damage classification performance; therefore, it should be overcome with approaches that handle multimodality and heterogeneity of the data as nonlinear models or clusters.

Finally, in this study, it is assumed that the bridge operates within its undamaged condition, under operational and environmental variability from 1 to 3470 observations. On the other hand, the bridge is considered damaged from 3471 to 3932 observations. Nevertheless, for generalization purposes, the feature vectors were split into the training and test matrices. As shown in Figure 7, the training matrix, $\mathbf{X}^{3123\times4}$, is composed of 90%

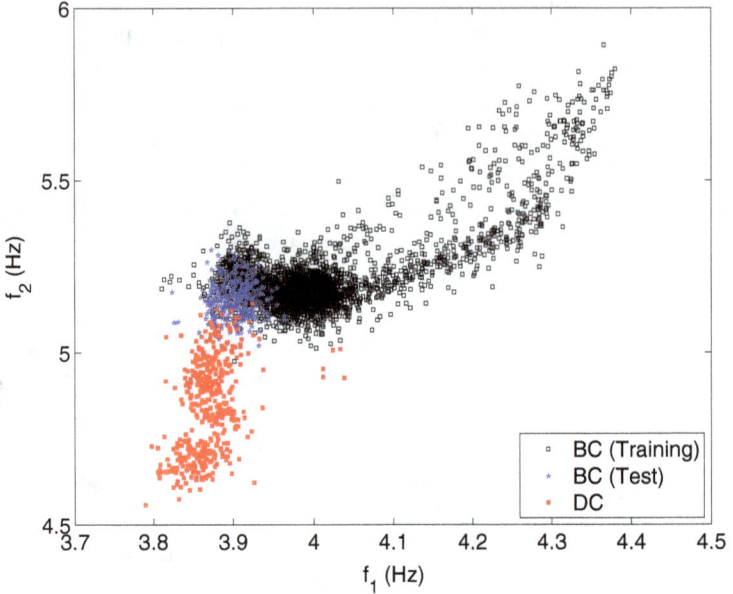

Figure 8.   Feature distribution of the two most relevant natural frequencies.

of the feature vectors from the undamaged condition. The remaining 10% of the feature vectors are used during the test phase to make sure that the DIs do not fire off before the damage starts. The test matrix, $\mathbf{Z}^{3932 \times 4}$, is composed of all the data sets, even the ones used during the training phase.

4.2. Damage detection on data sets from the Z-24 Bridge

Herein, the machine learning algorithms introduced in Section 3 are applied on the damage-sensitive features detailed above. The parameters required for the MSD, GMM, and AANN were defined following the recommendations described in Ref. 32. In the case of the PCA and KPCA, the parameters were computed based on criteria explored in Ref. 60. For the bioinspired algorithms, the parameters were chosen as indicated in Refs. 48 and 49. In addition, for all algorithms, the threshold for damage classification is defined based on the 95% cut-off value over the training data. The classification performance of the algorithms are summarized in Table 3.

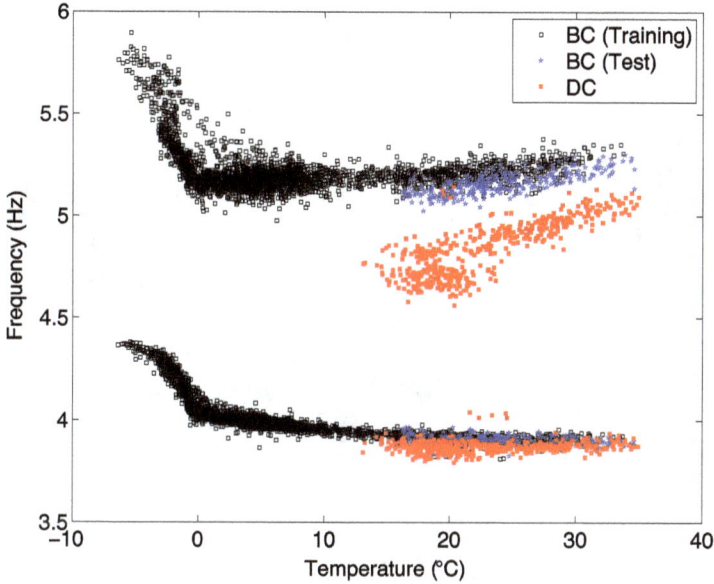

Figure 9. First two natural frequencies as a function of the ambient temperature.

Table 3. Damage detection performance for each algorithm.

Algorithm	Type I errors	Type II errors
MSD	162	190
GMM[a]	165.45 ± 1.15	8.25 ± 1.89
PCA	161	143
KPCA	172	4
AANN	174	6
GEM-GA[a]	166.00 ± 0.00	6.00 ± 0.00
GEM-PSO[a]	166.05 ± 0.22	6.00 ± 0.00

Note: [a]Damage classification performance (average ± standard deviation) for 20 executions.

A first attempt for data normalization and damage detection using the MSD and PCA algorithms is shown in Figure 10. The intrinsic linear formulation of the techniques impacts in a poor data normalization and, consequently, in an unacceptable damage detection performance (respectively, 190 and 143 Type II errors), as shown in Table 3. The high peak presented during the baseline condition demonstrates that the data normalization step performed, by both algorithms, cannot eliminate the high influence of the temperature on the natural frequencies. Since the

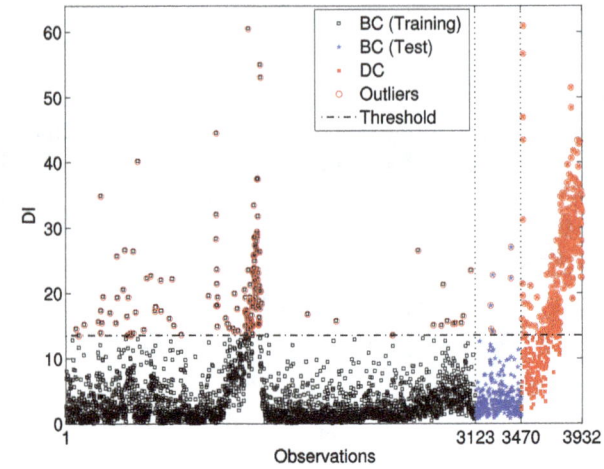

(a)

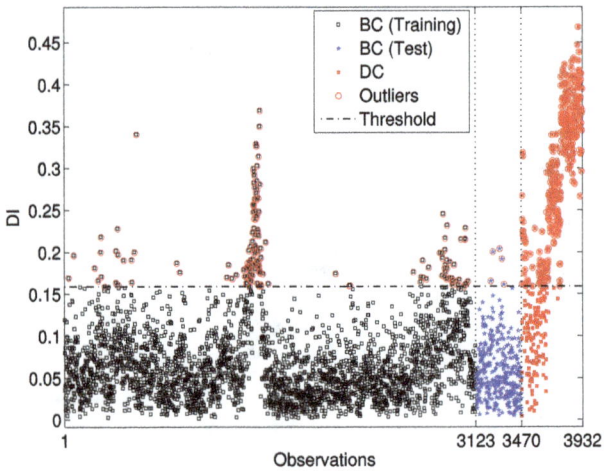

(b)

Figure 10.   Outlier detection based on the MSD (a) and PCA (b) algorithms.

environmental variability is not treated properly, several damaged cases are misclassified, which implicates a high risk to life-safety in real-world SHM applications.

A better damage detection performance is achieved when nonlinear algorithms are used to model the nonlinear relationship inherent in the

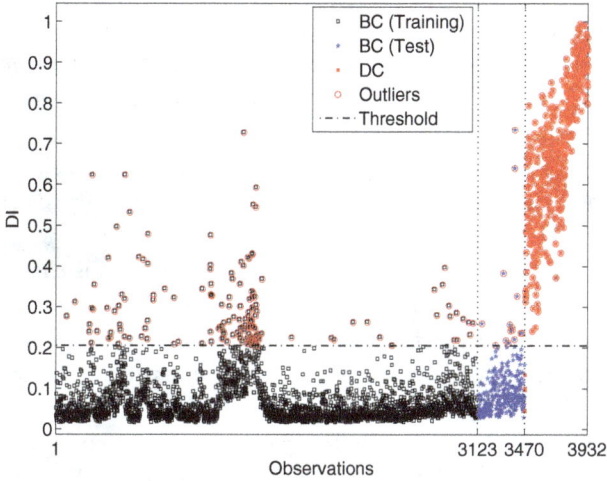

(a)

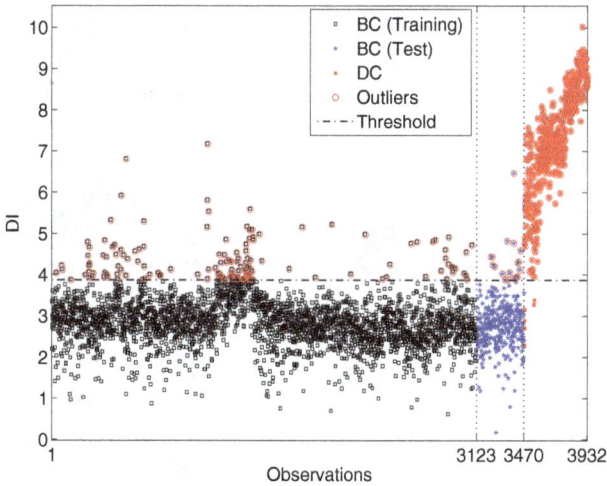

(b)

Figure 11.   Outlier detection based on the KPCA (a) and AANN (b) algorithms.

damage-sensitive features, as presented in Figure 11. By mapping the training observations, via a nonlinear kernel, to a high-dimensional feature space, the KPCA establishes a data model by taking into account that the appropriate dimensionality needs to be large enough to account for all normal condition but also small enough to be as sensitive as possible to damage. In a similar manner, the AANN applies a neural network to filter the environmental variability. In practice, based on Table 3, both techniques have almost the same performance, presenting acceptable damage detection rates (4 and 6 Type II errors, respectively) and moderate generalization of the learned model to the undamaged observations not used during the training phase (due to some loss of information in the dimensionality reduction performed by the KPCA and in the bottleneck layer selected by the AANN).

In an alternative manner, when the normal condition is characterized as data clusters (data normalization step), without any loss of information, the number of undamaged cases correctly classified increases, as demonstrated in Figure 12 for the GMM and GEM-GA algorithms and in Table 3 based on the lower number of Type I errors than the KPCA and AANN. However, if more executions (with different initial parameters) are considered for both techniques, the damage detection performance of the GMM deteriorates, as the algorithm is supported by an unstable method that depends on the choice of initial parameters. On the other hand, the damage classification performance of the GEM-GA (or GEM-PSO), which avoids dependency of the initial parameters, demonstrates high stability and reliability, in terms of reproducibility, based on the low standard deviation of the Type I and II errors (see Table 3).

Besides, the instability of the GMM algorithm is also related to the interchangeable number of components or clusters ($Q = 6.65 \pm 0.59$) present in each execution. In contrast, the GEM-GA (or GEM-PSO) maintains the same number of components ($Q = 7$) in all executions, which increases the robustness of the MA to overcome the sensitivity issue caused by the initial parameters.

In an overall analysis, except of the MSD and PCA, all algorithms attempt to maintain a monotonic relationship between the level of damage and the amplitude of the DIs, even when operational and environmental variability is present. This monotonic relationship reveals the damage progressive testing period, which indicates cumulative damage on long-term monitoring.

Finally, based on Table 3 and considering the Type I and the Type II errors, the best damage detection performance is reached by the GEM-GA and GEM-PSO, followed by the KPCA and AANN, where the former has the best performance in terms of Type II errors.

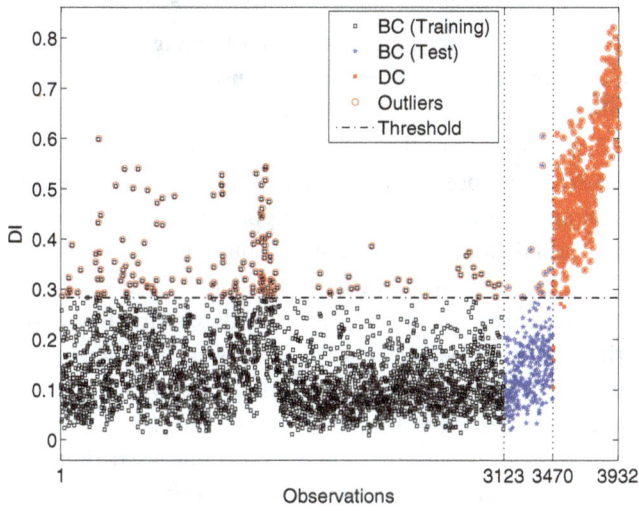

(a)

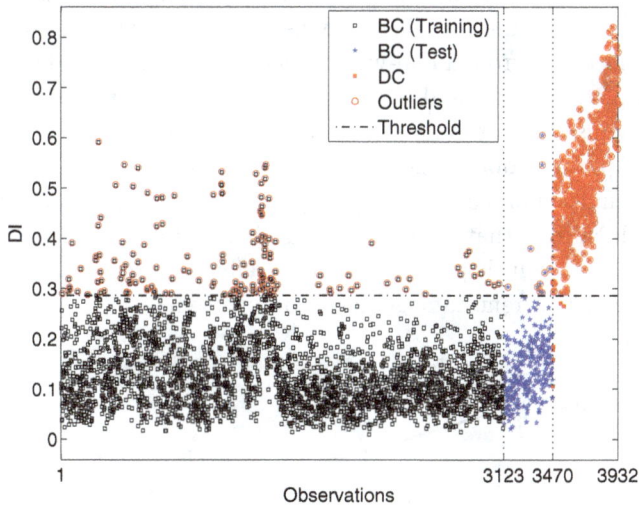

(b)

Figure 12.   Outlier detection based on the GMM (a) and GEM-GA (b) algorithms.

5. Summary, Main Conclusions, and Future Trends

In any real-world structure, the separation of changes in damage-sensitive features caused by damage from those caused by changing operational and environmental conditions is one of the biggest challenges to transit the SHM technology from research to practice. To address this issue, the SHM process was posed in the context of the SPR paradigm, where machine learning algorithms have an important role, as they are trained with past measured data, resembling the human brain. As long as these algorithms are fed with monitoring data collected during all the operational scenarios, they should be able to detect anomalies or patterns unseen during normal operation, without the need to measure the operational and environmental variations such as temperature and humidity. In the hierarchical structure of damage identification, this study was motivated by the need for robust incipient vibration-based damage detection methods. Thereby, it is mainly concerned with detection of damage in the structures.

Recognizing the applicability of the SPR paradigm for SHM, statistical methods from the machine learning field were tested and compared on data sets measured from a real-world structure: the Z-24 Bridge. From this comparison, the bioinspired algorithms proved to be promising methods that can be used to estimate the main state conditions of a structure and then to map a new observation into a DI. Unlike the other approaches, in terms of definition of main states, the bioinspired ones consists of two main parts: a global one that conducts the global search in the feature space; and a local one, which performs a more refined search around candidate solutions of the current damage detection problem.

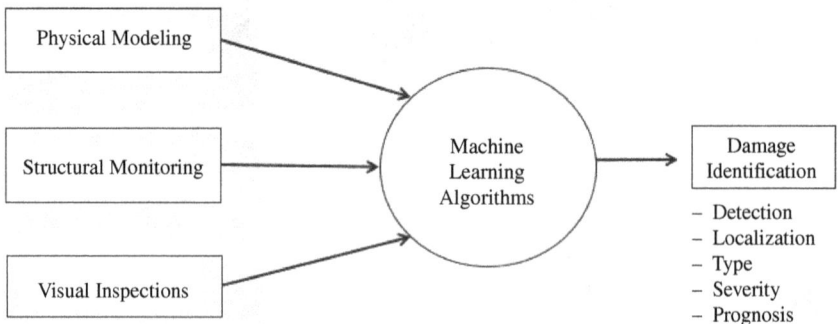

Figure 13. A new holistic pattern recognition paradigm to comprise physical modeling, structural monitoring, and information from visual inspections.

In a Big Data era faced currently, the machine learning algorithms also have the potential to simplify massive and complex monitoring data into simple damage indices and graphical representations. Depending on the formulation, some of these algorithms are able to filter linear and non-linear patterns and, therefore, enhance the characterization of structural parameters and behaviors.

In terms of future development, a holistic approach should be considered — holistic pattern recognition paradigm — which takes into account physical modeling, structural monitoring, and information from visual inspections, as represented in Figure 13. In this data fusion approach, the machine learning algorithms can learn from physics-, data-, and visual-based sources, which potentially improves their knowledge from the structures to better identify damage at early stages.

References

1. Farrar, C. R. and Worden, K. An introduction to structural health monitoring. *Philosophical Transactions of the Royal Society A* **365**(1851), pp. 303–315 (2007).
2. Farrar, C. R., Doebling, S. W., and Nix, D. A. Vibration-based structural damage identification. *Philosophical Transactions of the Royal Society of London A: Mathematical, Physical and Engineering Sciences* **359**(1778), pp. 131–149 (2001a).
3. Sohn, H. Effects of environmental and operational variability on structural health monitoring. *Philosophical Transactions of the Royal Society: Mathematical, Physical & Engineering Sciences* **365**(1851), pp. 539–560 (2007).
4. Worden, K. and Manson, G. The application of machine learning to structural health monitoring. *Philosophical Transactions of the Royal Society A* **365**(1851), pp. 515–537 (2007).
5. Figueiredo, E., Park, G., Farrar, C. R., Worden, K., and Figueiras, J. Machine learning algorithms for damage detection under operational and environmental variability. *Structural Health Monitoring* **10**(6), pp. 559–572 (2011).
6. Torres-Arredondo, M. A., Tibaduiza, D. A., Mujica, L. E., Rodellar, J., and Fritzen, C.-P. Data-driven multivariate algorithms for damage detection and identification: Evaluation and comparison. *Structural Health Monitoring* **13**(1), pp. 19–32 (2014).
7. Santos, A., Figueiredo, E., Silva, M., Sales, C., and Costa, J. C. W. A. Machine learning algorithms for damage detection: Kernel-based approaches. *Journal of Sound and Vibration* **363**, pp. 584–599 (2016).
8. Carden, E. P. and Fanning, P. Vibration based condition monitoring: A review. *Structural Health Monitoring* **3**(4), pp. 355–377 (2004).
9. Figueiredo, E. *Damage Identification in Civil Engineering Infrastructure Under Operational and Environmental Conditions*. Ph.D. Thesis, Doctor

of Philosophy Dissertation in Civil Engineering, Faculty of Engineering, University of Porto (2010).

10. Kullaa, J. Distinguishing between sensor fault, structural damage, and environmental or operational effects in structural health monitoring. *Mechanical Systems and Signal Processing* **25**(8), pp. 2976–2989 (2011).

11. Reynders, E. System identification methods for (operational) modal analysis: Review and comparison. *Archives of Computational Methods in Engineering* **19**(1), pp. 51–124 (2012).

12. Glisic, B. and Inaudi, D. Development of method for in-service crack detection based on distributed fiber optic sensors. *Structural Health Monitoring* **11**(2), pp. 161–171 (2012).

13. Reynders, E. and Roeck, G. D. A local flexibility method for vibration-based damage localization and quantification. *Journal of Sound and Vibration* **329**(12), pp. 2367–2383 (2010).

14. Baptista, F. G., Filho, J. V., and Inman, D. J. Real-time multi-sensors measurement system with temperature effects compensation for impedance-based structural health monitoring. *Structural Health Monitoring* **11**(2), pp. 173–186 (2012).

15. Kessler, S. S., Spearing, S. M., and Soutis, C. Damage detection in composite materials using Lamb wave methods. *Smart Materials and Structures* **11**(2), pp. 269–278 (2002).

16. Gyuhae Park, C. R. F., Hoon, S., and Inman, D. J. Overview of piezoelectric impedance-based health monitoring and path forward. *Shock and Vibration Digest* **35**(6), pp. 451–463 (2003).

17. Ihn, J.-B. and Chang, F.-K. Detection and monitoring of hidden fatigue crack growth using a built-in piezoelectric sensor/actuator network: II. Validation using riveted joints and repair patches. *Smart Materials and Structures* **13**(3), pp. 621–630 (2004).

18. Sohn, H., Park, G., Wait, J. R., Limback, N. P., and Farrar, C. R. Wavelet-based active sensing for delamination detection in composite structures. *Smart Materials and Structures* **13**(1), pp. 153–160 (2004).

19. Figueiredo, E., Park, G., Figueiras, J., Farrar, C., and Worden, K. Structural health monitoring algorithm comparisons using standard datasets, LANL Technical Report LA-14393, Los Alamos National Laboratory, Los Alamos, New Mexico, USA, (2009).

20. Farrar, C. R. and Worden, K. *Structural Health Monitoring: A Machine Learning Perspective*, John Wiley & Sons, Inc., West Sussex, UK, (2013).

21. da Silva, S., Júnior, M. D., Junior, V. L., and Brennan, M. J. Structural damage detection by fuzzy clustering. *Mechanical Systems and Signal Processing* **22**(7), pp. 1636–1649 (2008).

22. Diez, A., Khoa, N. L. D., Alamdari, M. M., Wang, Y., Chen, F., and Runcie, P. A clustering approach for structural health monitoring on bridges. *Journal of Civil Structural Health Monitoring* **6**(3), pp. 429–445 (2016).

23. Dervilis, N., Worden, K., and Cross, E. On robust regression analysis as a means of exploring environmental and operational conditions for SHM data. *Journal of Sound and Vibration* **347**, pp. 279–296 (2015).

24. Holmes, G., Sartor, P., Reed, S., Southern, P., Worden, K., and Cross, E. Prediction of landing gear loads using machine learning techniques. *Structural Health Monitoring* **15**(5), pp. 568–582 (2016).

25. Bradley, A. P. The use of the area under the ROC curve in the evaluation of machine learning algorithms. *Pattern Recognition* **30**(7), pp. 1145–1159 (1997).

26. Farrar, C. R., Sohn, H., and Worden, K. Data normalization: A key to structural health monitoring. In *Proceedings of the 3th International Workshop on Structural Health Monitoring* DEStech Publications, Inc., Stanford, Palo Alto, CA, pp. 1229–1238 (2001b).

27. Toivola, J. and Hollmén, J. *Feature Extraction and Selection from Vibration Measurements for Structural Health Monitoring*, Springer Berlin Heidelberg, pp. 213–224 (2009).

28. Su, Z., Wang, X., Cheng, L., Yu, L., and Chen, Z. On selection of data fusion schemes for structural damage evaluation. *Structural Health Monitoring* **8**(3), pp. 223–241 (2009).

29. Jayawardhana, M., Zhu, X., Liyanapathirana, R., and Gunawardana, U. Compressive sensing for efficient health monitoring and effective damage detection of structures. *Mechanical Systems and Signal Processing* **84** (Part A), pp. 414–430 (2017).

30. Farrar, C. R. and Lieven, N. A. J. Damage prognosis: The future of structural health monitoring. *Philosophical Transactions of the Royal Society of London A: Mathematical, Physical and Engineering Sciences* **365**(1851), pp. 623–632 (2007).

31. Worden, K., Manson, G. and Fieller, N. R. J. Damage detection using outlier analysis. *Journal of Sound and Vibration* **229**(3), pp. 647–667 (2000).

32. Figueiredo, E. and Cross, E. Linear approaches to modeling nonlinearities in long-term monitoring of bridges. *Journal of Civil Structural Health Monitoring* **3**(3), pp. 187–194 (2013).

33. McLachlan, G. J. and Peel, D. *Finite Mixture Models*, John Wiley & Sons, Inc., Wiley Series in Probability and Statistics, New York NY, United States, (2000).

34. Dempster, A. P., Laird, N. M., and Rubin, D. B. Maximum likelihood from incomplete data via the EM algorithm. *Journal of the Royal Statistical Society, Series B (Methodological)* **39**(1), pp. 1–38 (1977).

35. Figueiredo, E., Radu, L., Worden, K., and Farrar, C. R. A Bayesian approach based on a Markov-chain Monte Carlo method for damage detection under unknown sources of variability. *Engineering Structures* **80**, pp. 1–10 (2014).

36. Box, G. E. P., Jenkins, G. M., and Reinsel, G. C. *Time Series Analysis: Forecasting and Control*, 4th edn. John Wiley & Sons, Inc., Hoboken NJ, United States, (2008).

37. Jolliffe, I. *Principal Component Analysis*, 2nd edn. Springer-Verlag, New York NY, United States, (2002).

38. Yan, A.-M., Kerschen, G., Boe, P. D., and Golinval, J.-C. Structural damage diagnosis under varying environmental conditions–Part I: A linear analysis. *Mechanical Systems and Signal Processing* **19**(4), pp. 847–864 (2005).

39. Schölkopf, B., Smola, A., and Müller, K.-R. Nonlinear component analysis as a kernel eigenvalue problem. *Neural Computation* **10**(5), pp. 1299–1319 (1998).
40. Boser, B. E., Guyon, I. M., and Vapnik, V. N. A training algorithm for optimal margin classifiers. In *Proceedings of the Fifth Annual Workshop on Computational Learning Theory* ACM, Pittsburgh PA, United States, pp. 144–152 (1992).
41. Mercer, J. Functions of positive and negative type, and their connection with the theory of integral equations. *Philosophical Transactions of the Royal Society of London A: Mathematical, Physical and Engineering Sciences* **209**(441–458) pp. 415–446 (1909).
42. Keerthi, S. S. and Lin, C.-J. Asymptotic behaviors of support vector machines with gaussian kernel. *Neural Computation* **15**(7), pp. 1667–1689 (2003).
43. Widjaja, D., Varon, C., Dorado, A., Suykens, J. A. K., and Huffel, S. V. Application of kernel principal component analysis for single-lead-ECG-derived respiration. *IEEE Transactions on Biomedical Engineering* **59**(4), pp. 1169–1176 (2012).
44. Peres-Neto, P. R., Jackson, D. A., and Somers, K. M. How many principal components? stopping rules for determining the number of non-trivial axes revisited. *Computational Statistics & Data Analysis* **49**(4), pp. 974–997 (2005).
45. Sohn, H., Worden, K., and Farrar, C. R. Statistical damage classification under changing environmental and operational conditions. *Journal of Intelligent Material Systems and Structures* **13**(9), pp. 561–574 (2002).
46. Hsu, T.-Y. and Loh, C.-H. Damage detection accommodating nonlinear environmental effects by nonlinear principal component analysis. *Structural Control and Health Monitoring* **17**(3), pp. 338–354 (2010).
47. Kramer, M. A. Nonlinear principal component analysis using autoassociative neural networks. *AIChE Journal* **37**(2), pp. 233–243 (1991).
48. Santos, A., Figueiredo, E., Silva, M., Santos, R., Sales, C., and Costa, J. C. W. A. Genetic-based EM algorithm to improve the robustness of Gaussian mixture models for damage detection in bridges. *Structural Control and Health Monitoring*, 24, e1886. (2017).
49. Santos, A., Silva, M., Santos, R., Figueiredo, E., Sales, C., and Costa, J. C. W. A. A global expectation-maximization based on memetic swarm optimization for structural damage detection. *Structural Health Monitoring* **15**(5), pp. 610–625 (2016).
50. Silva, M., Santos, A., Figueiredo, E., Santos, R., Sales, C., and Costa, J. C. W. A. A novel unsupervised approach based on a genetic algorithm for structural damage detection in bridges. *Engineering Applications of Artificial Intelligence* **52**, pp. 168–180 (2016).
51. Kullaa, J. Structural health monitoring under nonlinear environmental or operational influences. *Shock and Vibration* **2014**, pp. 1–9 (2014).
52. Moscato, P. and Cotta, C. A gentle introduction to memetic algorithms. In *Handbook of Metaheuristics*, Springer US, Boston, MA, United States, pp. 105–144 (2003).

53. Neri, F. and Cotta, C. Memetic algorithms and memetic computing optimization: A literature review. *Swarm and Evolutionary Computation* **2**, pp. 1–14 (2012).

54. Miller, B. L. and Goldberg, D. E. Genetic algorithms, tournament selection, and the effects of noise. *Complex Systems* **9**(3), pp. 193–212 (1995).

55. Back, T. and Schwefel, H.-P. Evolutionary computation: An overview. In *Proceedings of IEEE International Conference on Evolutionary Computation*, pp. 20–29 (1996).

56. Roeck, G. D. The state-of-the-art of damage detection by vibration monitoring: The SIMCES experience. *Structural Control and Health Monitoring* **10**(2), pp. 127–134 (2003).

57. Peeters, B. and Roeck, G. D. Reference-based stochastic subspace identification for output-only modal analysis. *Mechanical Systems and Signal Processing* **13**(6), pp. 855–878 (1999).

58. Peeters, B., Maeck, J., and Roeck, G. D. Vibration-based damage detection in civil engineering: Excitation sources and temperature effects. *Smart Materials and Structures* **10**(3), pp. 518–527 (2001).

59. Peeters, B. and Roeck, G. D. One-year monitoring of the Z24-Bridge: Environmental effects versus damage events. *Earthquake Engineering & Structural Dynamics* **30**(2), pp. 149–171 (2001).

60. Reynders, E., Wursten, G., and Roeck, G. D. Output-only structural health monitoring in changing environmental conditions by means of nonlinear system identification. *Structural Health Monitoring* **13**(1), pp. 82–93 (2014).

Chapter 2

Data-Driven Methods for Vibration-Based Monitoring Based on Singular Spectrum Analysis

Irina Trendafilova*, David Garcia and Hussein Al-Bugharbee

Department of Mechanical and Aerospace Engineering
University of Strathclyde, UK
*irina.trendafilova@strath.ac.uk

This chapter studies the application of data-driven methods and specifically principal component analysis (PCA) and singular spectrum analysis (SSA) for purposes of damage assessment in structures and machinery. In this study, data analysis methods PCA and SSA are applied to the measured vibration signals in order to extract information about the state of the structure/machinery and the presence of a fault in it. Two applications are offered, one for damage assessment on a wind turbine blade and another one for fault diagnosis in rolling element bearings. The results demonstrate strong capabilities of the investigated methodology for both structural damage detection and rolling element fault diagnosis. Eventually, a discussion about the capabilities of the studied methodology and the way forward regarding extending its capabilities and applications is offered.

Keywords: VSHM; Singular spectrum analysis (SSA); Principal component analysis (PCA); Outlier principle; Structural and machinery monitoring; Rolling element fault detection; Wind turbine blade.

1. Introduction

In this section, the main idea of vibration-based monitoring as applied to structures and machinery will be discussed. Then, data-driven methods will be introduced and their place among the vibration-monitoring (VM) methods will be discussed. Later, the focus will be placed on principal component analysis (PCA) and singular spectrum analysis (SSA) and the way these can be applied to structural and machinery damage and fault assessment.

Structural health monitoring (SHM) plays an important role in the inspection of the structural integrity of most advanced materials. Most structures are subjected to vibrations, and therefore vibration-based SHM (VSHM) methods present an attractive possibility for monitoring.

Analogically, fault inspection and monitoring is an integral part for most complex machinery. Most machines vibrate during their operation and for them VM is one of the most widely used monitoring methods.

The background of VSHM and VM is in the fact that any change introduced within the structure/machine result in changes in its vibration response. Accordingly, a fault or damage within the structure/machine will incur changes in the vibration response no matter whether one is looking at the free or a forced vibration response. The important characteristics of VSHM and VM methods are that they are global and as such can be used to inspect parts which are difficult or impossible to access and also can be applied when the location of the damage fault is not known in advance.

In general there are two types of vibration-based monitoring methods depending on their nature: model-based ones and non-model based or data-driven methods.[1] The first type of methods rely on a model for the structure/machinery inspected and generally compare the modeled and the recorded vibration responses and, on the basis of this, extract information about the presence of a fault/damage, its location and/or its extent. Currently and historically, most structural and machine dynamic models assume some level of linearity, some models are even totally linear in the sense that they assume linear dynamic as well as linear material/structural behaviors. Unfortunately, most machines and structures demonstrate quite well-expressed nonlinear behavior as result of nonlinearities due to: material (e.g. in the case of composite materials), boundary conditions or nonlinearities that come from the structure and the connections. On some occasions even though the structure or the machine demonstrates nonlinear dynamic behavior, the linear model can still present a good approximation. But, in a number of cases, especially for cases of strong/distributed nonlinearities (as is the case for composite materials and/or complex machinery), the nonlinearities cannot be neglected, and hence the linear approximation cannot be considered as a good alternative. Using nonlinear modeling and applying it further for damage assessment is a rather complicated task and it has its challenges and limitations. On a number of occasions even though the model attempts to take into account the nonlinearities of the structure, the modeled response might still be different from the measured one. Then, an updating process might regard these differences as damage/fault-caused ones. On the contrary, *data-driven* methods do not assume any model or linearity. They just regard the measured signals as data and use certain data transformations in order to extract information from this data. The data-driven methods have become quite popular recently for the purposes of structural and machine VM.[1,2] They have demonstrated their capabilities for addressing the problems for damage detection and identification.[3] Most data-driven methods eventually classify the measured data to two or more categories (e.g. healthy and damaged).

Most data-driven structural and machinery diagnosis methods include three main stages: Stage 1, *Data acquisition*, Stage 2, *Signal analysis*, and Stage 3, *Diagnosis. Data acquisition* is the process of collecting useful information (i.e. signal) about the system. In this step, the number, type, locations, and sensitivity of the sensors to collect the signal are determined. The *signal analysis* step includes subjecting the signal to certain pretreatment and extracting some compact information, which might be in the form of "features" which are further used for monitoring the health status of the structure/machinery or for distinguishing among different health conditions. *Diagnosis* is the step in which the structure/machinery is assigned a category corresponding to one of the possible health conditions (e.g. healthy and damaged/faulty).

There are two main types of data-driven methods according to the type of classification procedure that they use for the purposes of damage assessment. The first type of methods are those that only use the data from the healthy state for the purposes of damage diagnosis, and these are the so-called unsupervised methods. The second type of methods utilize data from the healthy and the damaged states and make the decision on the basis of these two or more data categories; these are referred to as supervised methods. The first type of methods generally use the outlier principle in order to detect damage. Such methods are generally only capable of damage detection (level 1) and they need additional procedures in order to distinguish between different data categories stemming from different damage/fault types or sizes. The supervised type of methods are generally capable to do the higher stages of a monitoring method in terms of diagnosis.

Depending on the level of diagnosis, the classification might distinguish between two groups only (healthy and faulty) or can be extended to classify the structural/machinery condition to more than two categories (like e.g. corresponding to the fault size or the fault type). Many studies use classification/categorization process to perform the diagnosis, and many others apply pattern recognition for the purposes of detection and diagnosis.[3] The simplest way of distinguishing between two or sometimes more categories is to use a threshold-based method. Then, a certain variable/feature is measured and if it is below the threshold, the structure/machinery is considered healthy. Once the feature exceeds the threshold, the structure/machine passes into faulty/damaged state. This can be extended to more than two categories if multiple thresholds are used that might correspond to different fault/damage extensions.[4] Different pattern recognition techniques can be used for the purpose of recognition between two or more structural/machinery states. A lot of authors suggest neural networks-based classifiers for this purpose.[5] For the signal analysis stage, a number of transformations can be used, which range from the simplest ones that estimate some statistical moments to much more complicated ones that use, e.g. regressive and autoregressive modeling.[6] These include, e.g. extracting some statistical features and assessing their values at different structural/machinery conditions. Kurtosis and crest factor are two of the most commonly used features in machinery fault diagnosis problems. Both these factors characterize the "peakiness" of a signal. Some studies basically use the frequency domain vibration signal where the repetitive signal components will correspond to peaks at the frequency of repetition. Furthermore, the signal can be divided into different frequency bands, and/or some frequency bands may be filtered depending on the frequency ranges of interest.[7]

The data-driven methods usually use time series analysis methods in order to transform the measured data and extract information for the presence of a fault/damage and for its type and size.[8,9] A time series can be defined as a sequence of measurements of a time-dependent variable, e.g. acceleration or velocity, collected over a number of discrete time points. The concept of time series analysis is very popular in climate and financial research fields[10] and a large number of techniques have been developed for purposes of analysis or predicting of the future values of time series. These techniques take into account some aspects of the internal structure in the data. Some of these techniques offer representing a time series by parametric models such as autoregressive (AR) modeling.[11] Others, such as the SSA[12] are used to decompose time series into a number of independent components that can have some meaningful interpretation such as trend and periodic components.

This study describes two methods that were developed for two different purposes, for SHM and for machinery fault diagnosis. In the latter case, the machinery faults to be identified are rolling element bearing faults. Both methods are data-driven and both use SSA in different forms, as part of the methodologies, for structural and machinery fault/damage diagnosis.

2. PCA and SSA and Their Application for the Purposes of Structural and Machinery Monitoring

PCA is a data analysis technique that has been quite extensively used for the analysis of climatological, medical, and financial data. It found popularity recently in engineering and especially for the purposes of structural and machinery monitoring. The main idea of PCA is to reduce the data dimensionality while at the same time retaining most of its variance.[13] It has some attractive properties which are beneficial for applications within the field of vibration-based monitoring. PCA decomposes the original data into a number of independent components, the first several of which contain most of the data variance. For the case of data from multiple categories, the new variables, the principal components (PCs), tend to be grouped according to these categories, as PCA tends to reduce the distance between data from the same category while at the same time increasing the distance between data from different categories. These properties make PCA useful for the analysis of data coming from e.g. different damage states. There are a number of papers that suggest the application of PCA for structural and machinery monitoring purposes.[14–17] Several investigations suggest the selection of certain features from the time or frequency domain of the vibration response signals, which can be considered as independent, and subjecting them to PCA.[18] On most occasions, these are certain frequency components which, e.g. correspond to peaks in the spectrum or time components that are far enough from each other to be considered independent. Some studies use the natural frequencies as initial data and decompose them into PCs.[19] In some papers, PCA is applied twice — once on the initial data and a second time on the decomposed data.[20] The authors suggest that this is useful for excluding the variability due to the environment and the operational conditions. As the first several PCs retain a high proportion of the variance, some authors argue that they also retain most of the useful information about the structure, embedded in the data analyzed, and accordingly they suggest the use of the few first principle components for the purposes of vibration analysis and vibration-based damage assessment. Other investigations suggest that the higher PCs containing the smallest amount of variance will contain information about the presence of damage, and accordingly suggest to use these in order to form damage features.[21] These studies propose that the first PCs are mostly responsible for the noise within the data rather than its internal structure and thus conclude that the last PCs are likely to contain most of the useful information about the structural/machinery dynamics.

SSA is a variation of PCA which has been developed especially for time series analysis. PCA regards the initial data as statistically independent. For the case of time series, two or more consecutive time series, components are not independent on most occasions as there is common information contained in these components, i.e. they contain "mutual information". SSA is designed to deal with non-independent data. SSA can be applied in the time and the frequency domain.[13] The aim of SSA is to decompose the original signal using a small number of independent and more interpretable components which can be used for trend identification, detection of oscillatory components, periodicity extraction, signal smoothing, noise reduction, feature extraction, and detection of structural changes in the time series. SSA has been applied for diverse applications ranging from weather forecasting[22] and financial mathematics[12] to historical[23] and economical time series where the signals are highly non-stationary with no signs of periodicities.[24]

SSA considers all rotational patterns contained in the vibration response rather than those corresponding to a particular frequency. For this reason, when SSA is applied to a time domain vibration signal, the signal is decomposed into harmonic and non-harmonic oscillatory components. Hence, SSA can be regarded as a kind of nonlinear spectral analysis.[25] This is why it can be argued that close modes, which are a feature of many nonlinear dynamic systems, will not be lost when SSA is used for decomposition purposes. There are a small number of publications related to the application of SSA for structural vibration analysis and for VSHM. In Ref. 26, SSA is applied for structural monitoring and damage diagnosis in bridges by using an eigenvalue ratio difference between the first two eigenvalues. Thus, when the difference between the first two eigenvalues increases, it can be considered as an indication for the occurrence of an irregularity. In the same study, the residual errors were measured by comparing the reconstructed vibration responses based on the SSA decomposition with the measured ones. In Ref. 27, the performance of an SSA-based methodology was compared to another method covariance-driven stochastic subspace identification (SSI-COV), and the authors claim the superiority of the SSA-based method in terms of speed and precision.

SSA has been applied in a couple of studies featuring machinery vibration-based fault diagnosis.[16,28–30] Bubathai[30] published the first study which uses SSA for classifying rolling element bearing signals as healthy and faulty (with a fault on inner raceway) for detection purposes only. In this study, the vibration acceleration signals from both categories were subjected to SSA and the original signals are decomposed into two PCs: trend and residuals. Then, only the trend component was considered for further analysis. A number of statistical features such as peak value and the standard deviation are obtained from the trend. These features are further used to form the feature vectors (FVs) which are eventually used as input for a neural network classifier. SSA was used as a multi-decomposition analysis technique.[16] In this study the number of singular values, which preserves a specific predetermined variance percentage, is used as an indicator for fault presence. Two different feature sets were obtained from the application of SSA[17] and used as FVs. The first FV was made of the singular values and the second one- from the energy of the first time domain PCs. The bearing condition classification was performed by using those FVs as input to a back propagation neural network classifier.

This chapter considers the application of SSA for two different purposes, namely, for SHM purposes as applied for delamination detection in wind turbine blades and for the purposes of machinery fault diagnosis as applied for rolling element fault identification.

The first application for SHM purposes decomposes the frequency domain signal rather than the measured time series. The argument for this is that a frequency domain representation presents the oscillatory patterns contained in the vibration response in more interpretable and ordered manner. It uses all the steps of SSA — decomposes and after that reconstructs the initial signal. The reconstructed signal demonstrates very good agreement with the original one, from which one can conclude that the used decomposition has preserved most of the important signal characteristics and the information contained within the signal by using a reduced number of reconstructive components. When the methodology is performed in the frequency domain, it can be observed that the first reconstructive component contains the general trend of the spectral line and the rest of reconstructive components contain the fluctuations along the spectral line. Therefore, only reduced number of reconstructive components is needed to describe the general behavior of the spectral line.[31-32] Most authors who suggest application of SSA apply this to the original time domain signal.[33]

The second application is for rolling element fault diagnosis. In this application, the time domain signal representation is used. This application suggests a classification process in order to first detect the presence of a fault and then identify the fault type and eventually estimate the fault size. Very little rolling element fault assessment studies offer a complete identification in terms of detection, type assessment, and extent estimation. This is one of the main contributions of this second application of SSA. The whole process suggested is very easy and can be made automatic so that it can be used for practical fault identification. This identification process uses only the first stage of SSA — the signal decomposition into PCs in order to extract fault features and assess the fault. This is the second important advantage of the suggested methodology for machinery diagnosis. It uses the signals corresponding to the baseline condition and decomposes them into PCs. A number of the first PCs are chosen, according to the percentage of variance they contain, and a baseline space is created using these first several PCs. Subsequently any of the signal that have to be identified are only projected on the baseline space in order to be compared to the healthy state. Most methods that use SSA for structural and machinery monitoring purposes use all the three stages of the process — the embedding, the decomposition, and the reconstruction stage. Thus, it should be noted that the method presented here is actually much simpler as it contains less computation, which makes it quite easy for application purposes. The SSA decomposition divides the data into categories which are well distinguishable and can be assigned to the different fault types and fault sizes. It should also be mentioned that the methods developed and described in this study give a very good and correct classification rate. The rolling element fault diagnosis method was compared in terms of performance to some other methods that were applied for the same data and it is observed that it outperforms them.

The next paragraph introduces the two methods — the one for SHM and the method for bearing fault diagnosis. The following two paragraphs are dedicated to two particular applications of the introduced methodologies. They introduce the case studies, the experiments performed, the data collected, as well as the transforms applied. They also introduce some results from both applications and offer a short analysis of these results. The last section of this chapter suggests a discussion of the developed methods and the results obtained for the two case studies.

3. Two SSA-Based Damage Assessment Methodologies for Structural and Machinery Monitoring

This section introduces two damage assessment methodologies as developed for SHM and for rolling element fault diagnosis. As mentioned before, both methods are based on SSA decomposition, but they use different modifications of SSA and are applied in different ways. The first method is applied for SHM and it uses the signal in the frequency domain. The second method is developed for rolling element bearing diagnosis and it decomposes the time domain signal. The rationale behind this is that for the first case, most of the information regarding the vibration modes of the structure needed to be retained, while in the second case it is assumed that most of the information about the system can be recovered using the lagged time signals. In this paragraph, the basic steps of SSA are first introduced and then the two methods for damage assessment as applied for different purposes are presented.

3.1. *SSA basic steps*

SSA, as a decomposition method, has several major stages, namely, data collection, embedding, decomposition, and reconstruction.[34] The method for structural damage assessment considered here uses all these three steps. The method for bearing fault diagnosis discussed below does not follow the reconstruction stage.[35] The latter creates a reference space related to the healthy state based on the decomposed components only.

3.1.1. *Data collection*

In general, SSA as well as PCA are data analysis methods and they assume multiple realizations of the signal in their analysis process, which might be taken in different locations or just as a result of a number of measurements. In this study, the data collection is done in different ways for the SHM process, and for machinery monitoring procedure. In the SHM process, the acceleration signal is measured in a number of different positions. In the second application, a long enough signal is measured and it is later on divided into segments which are treated as different realizations/measurements. For the purposes of data collection, the measured signals are initially arranged into a matrix as columns.

3.1.2. *Embedding*

The embedding stage in SSA is used to embed a lagged version of the measured signal into a matrix, thus expanding the information contained in the signal into more dimensions according to Takens embedding theorem.[36] Embedding can be done in different ways, but in all cases a window of certain size, smaller than the original size of the signal, is used and each signal vector is embedded (expanded) into a matrix using lagged versions of the vector itself. Generally speaking, as a result of the embedding each signal/vector is transformed/unfolded into a matrix. Usually, the embedding is performed in the time domain, but it might be done in the frequency domain as well, if SSA is applied on the signal spectra rather than on the time signature. Eventually, the embedding is done for all the data vector realizations, and the corresponding matrices are put together to form the final data matrix, also called an embedding matrix, which is subjected to the second step, i.e. the PCs decomposition. In this study, two different embedding procedures are used for the development of the two methods, as is presented in the following sections. One of the methods is applied on the frequency domain signal representation and the other one on the original time domain.

3.1.3. *Decomposition*

The decomposition into PCs is performed using the covariance matrix of the embedding matrix. The covariance matrix gives the covariance between the signal realizations. This covariance matrix is then subjected to eigenvalue decomposition to obtain its eigenvalues and eigenvectors. Each eigenvalue represents the partial variance of the original time series in the direction of the corresponding eigenvector. Projecting the embedding matrix onto each eigenvector provides the corresponding PCs.

3.1.4. *Reconstruction*

The idea of this stage is to reconstruct the original signal using a linear combination of the obtained PCs. The original signals can be reconstructed by a linear combination of all or just a few of the PCs. Different criteria can be used to select the number of PCs to be used in the reconstruction process. As a result, the so-called reconstructed components (RCs) are obtained. Using a number of or all the reconstructed components, the original signal can be reconstructed with certain accuracy. A good agreement between the original and the reconstructed signal (small difference error) means that the decomposition process was able to capture most of the information contained in the original data.

3.2. *Development of an SSA-based technique as a methodology for VSHM*

This method follows the general steps of SSA as explained above, which are the data collection, the embedding, the decomposition into PCs, and the reconstruction. The main stages involved in the diagnosis are: (1) creation of reference space relative to which all the new signals will be compared and (2) damage detection. The reference space is created using SSA. The recognition between healthy and faulty structures is done by comparing each set of new data to the baseline/healthy set of data which is transformed into new coordinates in order to create the baseline/reference space. A new signal is then projected onto the reference space and is classified as healthy or damaged based on a predefined threshold.

In the following sections, the main steps of the method are briefly described.

3.2.1. *Data collection*

A discrete acceleration signal is measured in N time sampling points. Each signal is standardized to have a zero mean and unit variance. As mentioned above, one is in possession of a number $m = 1, \ldots, M$ of realizations of such acceleration signals, each of which can be represented by a signal vector as shown in Equation (1):

$$\mathbf{x}_m = (x_{1,m}, x_{2,m}, \ldots, x_{N,m}) \tag{1}$$

Each signal vector $\mathbf{x}_m$ is then transformed to the frequency domain to obtain a new signal vector $\mathbf{y}_m$ with a length $\bar{N} = N/2$. All the obtained frequency domain signal vectors are stored into a matrix $\mathbf{Y}$ as shown in Equation (2).

$$\mathbf{Y} = (\mathbf{y}_1, \mathbf{y}_2, \ldots, \mathbf{y}_M) \tag{2}$$

3.2.2. *Creation of the reference space*

A reference space is built using the signals, measured on the healthy/pristine structure. The steps for creating the reference space are: Embedding, decomposition, and reconstruction. Each of these steps is briefly described below.

3.2.2.1. Embedding

This step creates an embedded matrix of the signal vector $\mathbf{y}_m$. Each signal vector $\mathbf{y}_m$ is embedded into a matrix $\tilde{\mathbf{Y}}_m$ by using L lagged copies of the signal/vector itself as shown in Equation (3), where L is the sliding window size.

$$\tilde{\mathbf{Y}}_m = \begin{pmatrix} y_{1,m} & y_{2,m} & y_{3,m} & \cdots & y_{L,m} \\ y_{2,m} & y_{3,m} & y_{4,m} & \cdots & y_{(L+1),m} \\ y_{3,m} & y_{4,m} & y_{5,m} & \cdots & \vdots \\ y_{4,m} & y_{5,m} & \vdots & \cdots & \vdots \\ y_{5,m} & \vdots & \vdots & \cdots & y_{\bar{N},m} \\ \vdots & \vdots & y_{(\bar{N}-1),m} & \cdots & 0 \\ \vdots & y_{(\bar{N}-1),m} & y_{\bar{N},m} & \cdots & 0 \\ y_{(\bar{N}-1),m} & y_{\bar{N},m} & 0 & \cdots & 0 \\ y_{\bar{N},m} & 0 & 0 & \cdots & 0 \end{pmatrix} \tag{3}$$

The embedding process is applied to each signal/vector and then all $\tilde{\mathbf{Y}}_m$ matrices are used to create the data-embedded matrix $\tilde{\mathbf{Y}}$ given by Equation (4). The dimension of the matrix $\tilde{\mathbf{Y}}$ is $[\bar{N} \times (ML)]$. The sliding window size L is usually selected so that $M < L$ and $L \leq \bar{N}/2$.

$$\tilde{\mathbf{Y}} = (\tilde{\mathbf{Y}}_1, \tilde{\mathbf{Y}}_2, \ldots, \tilde{\mathbf{Y}}_M) \tag{4}$$

3.2.2.2. Decomposition into PCs

The embedding matrix $\tilde{\mathbf{Y}}$ obtained above (see Equation (4)) is decomposed into PCs. First, the covariance matrix $\mathbf{C_Y}$ of the matrix $\tilde{Y}$ is calculated according to Equation (5).

$$C_Y = \frac{\tilde{\mathbf{Y}}^t \tilde{\mathbf{Y}}}{\bar{N}} \tag{5}$$

where $\tilde{\mathbf{Y}}^t$ is the transpose matrix of $\tilde{\mathbf{Y}}$. The covariance matrix $\mathbf{C}_Y$ has a dimension $[(ML) \times (ML)]$ and gives the covariance between signal realizations. The matrix $\mathbf{C}_Y$ is subsequently subjected to eigenvalue decomposition as shown in Equation (6):

$$\mathbf{E}_Y^t \mathbf{C}_Y \mathbf{E}_Y = \mathbf{\Lambda}_Y \tag{6}$$

The matrix $\mathbf{\Lambda}_Y$ is a diagonal matrix with eigenvalues λ_k on its diagonal in decreasing order and $\mathbf{E}_Y$ contains all eigenvectors $\mathbf{E}^k$ in columns in the same order as the corresponding eigenvalues. Each eigenvector $\mathbf{E}^k$ is composed of M consecutive L-long segments, depending on the number of realizations and the sliding window size, respectively, with its elements denoted by $E_{m,l}^k$. Each PC $\mathbf{A}_k$ associated with each eigenvector $\mathbf{E}^k$ is a single-channel vector calculated by projecting the matrix $\tilde{\mathbf{Y}}$ onto $\mathbf{E}_Y$ as shown in Equation (7) where $n = 1, \ldots, \bar{N}$ (see Ref. 22).

$$A_n^k = \sum_{l=1}^{L} \sum_{m=1}^{M} Y_{m,n+l} E_{m,l}^k \tag{7}$$

Accordingly, it can be seen from Equation (7) that each PC contains characteristics from all the M signal vector realizations.

3.2.2.3. Reconstruction

For a given set of indices K, the RCs are calculated by convolving the PCs with the eigenvectors $\mathbf{E}_Y$, so that the kth RC at n-value for an m-realization is given by Equation (8) where $n = 1, \ldots, \bar{N}$ (see Ref. 22).

$$R_{m,n}^k = \frac{1}{L_n} \sum_{l=1}^{L} A_{n-l}^k E_{m,l}^k \tag{8}$$

Each $R_{m,n}^k$ value is normalized by a normalization factor L_n which is shown in Equation (9):

$$L_n = \begin{cases} n & 1 \leq n \leq L-1 \\ L & L \leq n \leq N \end{cases} \tag{9}$$

The matrix $\mathbf{R}$ includes all reconstructed components for all the original signal vectors in columns.

3.2.3. *Feature extraction*

A FV is obtained for each newly observed signal which will be subjected to the damage assessment process, by comparing its similarity to the reference space. The FV is calculated by multiplying/projecting the observed signal vector $\mathbf{y}$ to the reference space matrix $\mathbf{R}$ as shown in Equation (10) below where $j = 1, \ldots, L$.

$$T_j = \sum_{n=1}^{\bar{N}} y_n R_{n,j} \tag{10}$$

Then, the features T_j are arranged into a vector $\mathbf{T}$ with dimension L. The FV $\mathbf{T}$ characterizes the similarity of the observed signal $\mathbf{y}$ to the reconstructed reference space.

3.2.4. *Damage assessment*

The damage assessment in this study is done on the basis of a predetermined threshold.

First a baseline feature matrix $\mathbf{T_B}$, with a dimension $p \times s$ where p is the dimension of each FV $\{\mathbf{T}{:}p \leq L\}$ and s is the number of signal vectors used to define the baseline matrix, is created whose elements are obtained following Equation (11).

$$\boldsymbol{T}_B = \begin{pmatrix} T_{1,1} & T_{2,1} & \cdots & T_{p,1} \\ T_{1,2} & T_{2,2} & \cdots & T_{p,2} \\ \vdots & \vdots & \cdots & \vdots \\ T_{1,s} & T_{2,s} & \cdots & T_{p,s} \end{pmatrix} \tag{11}$$

The next step is to measure the similarity of an observed FV $\mathbf{T}^i = (T_{1,i}, T_{1,i}, \ldots, T_{p,i})$ to the baseline feature matrix $\mathbf{T_B}$ where i is the number of observation signal vectors considered. This is done on the basis of the Mahalanobis distance between the FV and the baseline matrix which is defined in Equation (12).

$$D_i = \sqrt{\left(\boldsymbol{T}^i - \boldsymbol{\mu}_B\right)^t \sum{}^{-1} \left(\boldsymbol{T}^i - \boldsymbol{\mu}_B\right)} \tag{12}$$

where $\boldsymbol{\mu}_B$ is the mean row of the baseline feature matrix $\mathbf{T}_B$ and $\boldsymbol{\Sigma}$ is its covariance matrix.

For the purposes of damage assessment, a threshold ϑ is defined and the above distance is compared to the threshold. If the distance to the baseline matrix is $D_i < \vartheta$, then the observed signal vectors are assigned to the baseline (healthy) category. If $D_i > \vartheta$, then the newly observed vector is considered to be outside the baseline category, i.e. it belongs to the damaged category.

3.3. An SSA-based technique for rolling element fault diagnosis

The method presented here was developed for the purposes of rolling element fault detection and diagnosis. This is a relatively simple methodology which also uses SSA for the purposes of decomposition of the measured vibration signals. In this case, the acceleration signals are measured on bearing housing and the time domain signals are used.

The methodology decomposes the signal vectors corresponding to the healthy bearing state using an SSA-based technique in order to build a baseline space. In this case, the data collection and the decomposition stages are only used and are only applied to a set of signals which are collected in the healthy/baseline condition of the bearings.

The eigenvectors obtained applying the SSA decomposition are used to calculate the PCs corresponding to the baseline/healthy condition of the bearings. A baseline space corresponding to healthy bearings' state is created. In this way, SSA is only used to transform the baseline signals. The new signals which have to be classified as healthy or damaged are not transformed, they are just projected onto the baseline space, and these projections are used for the purposes of fault detection.

The methodology consists of three basic phases, namely fault detection, fault type identification and fault severity estimation. In the fault detection phase, the signals are classified into two categories: baseline/healthy and non-baseline/faulty. In the second phase, the fault type is identified by assigning the signals to one of fault type categories: inner race fault (IRF) and outer race fault (ORF). In the consequent fault severity estimation phase, the severity of the fault is estimated by assigning the signals to categories corresponding to different fault severity levels.

The method can be divided into two main stages: building a baseline space and fault diagnosis. An SSA-based procedure is used to build the baseline/healthy space using the signals measured on the healthy bearings, and this is described in the next paragraph.

3.3.1. *Building the baseline space*

The baseline space construction uses an SSA procedure which is applied on the vibration signals $\mathbf{x}$ measured on the healthy bearings. It contains similar phases as the procedure described in the previous section, but only the data collection and the decomposition stages are used.

3.3.1.1. Data collection and embedding

The discrete acceleration signals are measured and collected in their time domain form so that each signal is represented as a vector $\mathbf{x}$ as introduced in Equation (1) (see Section 3.2.1).

Then, the embedding stage is performed to obtain the data embedding matrix $\tilde{\mathbf{X}}$ with dimension $(L \times K)$ where L is the sliding window size and $K = N - L + 1$, as shown in Equation (13).

$$\tilde{\mathbf{X}} = \begin{pmatrix} x_1 & x_2 & x_3 & \cdots & x_K \\ x_2 & x_3 & x_4 & \cdots & x_{K+1} \\ x_3 & x_4 & x_5 & \cdots & x_{K+2} \\ \vdots & \vdots & \vdots & \ddots & \vdots \\ x_L & x_{L+1} & x_{L+2} & \cdots & x_N \end{pmatrix} \tag{13}$$

3.3.1.2. Decomposition into PCs

At this stage, the covariance matrix of each embedding matrix $\tilde{\mathbf{X}}$ is obtained following Equation (14).

$$\mathbf{C}_X = \frac{\tilde{\mathbf{X}}^t \tilde{\mathbf{X}}}{K} \tag{14}$$

The covariance matrix $\mathbf{C}_X$ is subjected to singular value decomposition (SVD) according to Equation (15).

$$\mathbf{C}_X \mathbf{U}_k = \lambda_k \mathbf{U}_k \tag{15}$$

L eigenvalues λ_k and L corresponding eigenvectors which are ordered as columns of $\mathbf{U}_k$ are obtained in this manner. The eigenvalues λ_k are ordered in decreasing order and all eigenvectors in the columns of $\mathbf{U}_k$, are in the same order as their corresponding eigenvalues, so that the first several ones are responsible for a big part of the variance of the data. The PCs are obtained by projecting the embedding matrix onto the eigenvectors:

$$A_k^i(n) = \sum_{l=1}^{L} \tilde{X}^i (n + l - 1) U_k(l) \tag{16}$$

Therefore, the eigenvectors corresponding to the baseline/healthy state are arranged as columns of the matrix $\mathbf{U}$ as shown in Equation (17).

$$\boldsymbol{U} = (\boldsymbol{U}_1, \boldsymbol{U}_2, \ldots, \boldsymbol{U}_L) \tag{17}$$

The matrix $\mathbf{U}$ defines the baseline space. This is the full matrix made from all the eigenvectors. Usually, only a few of them are used for further purposes. There are several criteria mentioned in Ref. 16 that can be used for the selection of the number of eigenvectors. However, for the purposes of visualization, in our case the first three eigenvectors are used.

3.3.2. *Fault diagnosis process*

At this stage, new signals are presented and the aim is to diagnose the bearings on which they have been measured. This stage has two main phases: feature extraction and fault identification, which includes fault detection and fault type and size assessment. These are described in the following section.

3.3.2.1. Feature extraction

The fault identification here is done on the basis of certain features. So the first stage is to build these features using the measured signals. This is done by projecting the signal onto the baseline space defined by the matrix $\mathbf{U}$ (see Equations (17) and (18)).

For each new signal i, the embedding matrix $\tilde{\mathbf{X}}^i$ is calculated and then the PCs are obtained by projecting each $\tilde{\mathbf{X}}^i$ onto the $\mathbf{U}_k$ baseline space as shown in Equation (18).

The PCs obtained for each observation signal vector i are arranged into vector-column as shown in Equation (18).

$$\boldsymbol{A}^i = (\boldsymbol{A}_1, \boldsymbol{A}_2, \ldots, \boldsymbol{A}_L) \tag{18}$$

The FVs are calculated by the Euclidean norm of each PC as shown in Equation (19) where $p \leq L$ is the number of PCs considered.

$$T_p^i = \sum_{n=1}^{K} \left(A_p^i(n)\right)^2 \qquad (19)$$

Similarly, as in the previous methodology the feature values are arranged into a vector $\mathbf{T}$ with dimension $p \leq L$. The FV $\mathbf{T}$ characterizes the similarity of the observed signal $\mathbf{x}$ to the baseline space, which is based on the healthy bearing scenario.

3.3.2.2. Fault detection

The FVs obtained from the training sample corresponding to the healthy bearing condition are used to make the baseline feature matrix $\mathbf{T}_B$ as defined in Equation (13) (see Section 3.2.4).

The next step is to measure the similarity of an observed feature vector $\mathbf{T}^i = (T_{1,i}, T_{1,i}, T_{p,i})$ to the baseline feature matrix $\mathbf{T}_B$ where i is the number of observation signal vectors considered. This is done on the basis of the Mahalanobis distance of the vector $\mathbf{T}^i$ to the matrix $\mathbf{T}_B$ as defined in Equation (14) (see Section 3.2.4).

Similarly, as in Section 3.2.4., for the purposes of damage detection a threshold ϑ is defined and the obtained Mahalanobis distance D_i is compared with the threshold. If the distance to the baseline matrix is less than ϑ, $D_i < \vartheta$, then the observed signal vector is assigned to the baseline (healthy) category. If $D_i > \vartheta$, then the newly observed vector is considered to be outside the baseline category, i.e. it belongs to the damaged category.

3.3.2.3. Fault type and size identification

The next two steps of the diagnosis process are the determination of the fault type and its size estimation. They are done in the same way as the detection process. Training samples for each category corresponding to (1) different fault types and to (2) different fault sizes are built using a number of measured signals. Then each new signal, once converted into FV, is compared to these training samples in terms of its Mahalanobis distance. The signal/the FV is assigned to the category to which its Mahalanobis distance is the smallest one. That is, the 1 nearest neighbor (1NN) rule is used to determine the fault type and estimate its size. The fault type diagnosis is done first, and then within each type category, size categories training sets are created. Following the 1NN rule, the fault is assigned to the closest category in terms of its Mahalanobis distance.

3.4. *A case study for structural damage assessment in wind turbine blades*

3.4.1. *The case study*

This section applies the structural damage assessment methodology which was presented in Section 3.2 for a large SSP34 m wind turbine blade. The SSP34 m blade was mounted on a test rig at DTU Wind Energy facilities in Roskilde, Denmark. The data for this study (see Acknowledgement) was provided and belongs to Brüel & Kjær.

In this study, the main objective was to assess the artificially introduced damage in the trailing edge (TE) of the blade. The blade was excited by an electromechanical actuator to invoke a free-decay response. The aim of the analysis was to detect the damage when the blade was excited at different actuation locations. The experiment was performed for the case of sensors placed in different locations along the blade. In this section, the experiment is described and after that certain results that have been obtained are presented using the above described methodology.

The SSP34 m blade was mounted in a cantilever position in a test rig as shown in Figure 1. The blade was clamped at the root-end as it would be mounted on the rotor hub of the wind turbine. Twenty B&K tri axial accelerometers Type 4524-B were attached to the blade. 10 of these accelerometers were placed on the leading edge and the other 10 on the TE. The data measured in the direction perpendicular to the blade surface was analyzed. The placement of the accelerometers did not follow any systematic approach for selecting an optimum number and/or location. The excitation was provided by a signal generator, which was set to generate an amplified rectangular pulse fed to the actuator for each actuator hit. The free-decay response of the blade invoked by the applied rectangular pulse force was measured.

To perform the data-driven methodology, a full scale 34-m blade, manufactured by SSP Technology A/S, was mounted to a test rig under laboratory conditions (see Ref. 37 for more information). The debonding of one of the edges (TE or LE) between the top and bottom shell is a common damage which occurs in wind turbine blades. The damage was introduced artificially into the blade by drilling a series of holes through the glue between the shells

Figure 1.   The real-scale turbine blade tested.

of the blade on the TE. Then, using a saw the holes were merged into a crack which was opened by a chisel and a hammer. The crack was gradually extended up to 120 cm.

As mentioned previously, several different actuation positions and several different sensor placements were realized. The damage was also simulated in different locations along the blade. Figure 2 shows the positions of the sensors and actuators and the damage for the scenario tested for which the below results are presented.

3.4.2. *Damage assessment procedure*

The damage detection procedure was separately applied using the data from one sensor at a time. The reason for this is because one of our aims was to analyze the best sensor location for the process.

The reference states were created by the free-decay acceleration response sampled at 16,384 Hz measured for the pristine blade. The reference state was created using $M = 10$ signal vector realizations from the healthy blade and a sliding window size $L = 10$. The vibration responses

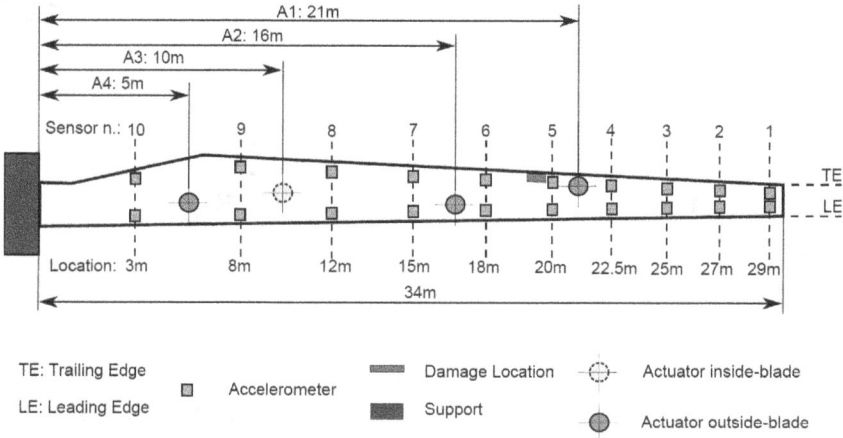

Figure 2. A schematic of the blade with the positions of the damage, the sensors, and the actuators.

were transformed into the frequency domain, and discretized into vectors of length $N = 2048$. With these signal vectors, the embedding matrix $\mathbf{Y}$ had a dimension 2048×100. The eigen-decomposition of the covariance matrix of the matrix $\mathbf{Y}$ yielded 100 eigenvalues and their corresponding eigenvectors. Therefore, the reference space matrix $\mathbf{R}$ (see Section 3.2.2) had a dimension 2048×10.

The FVs were obtained by projecting the observation signal onto the reference space. A dimension of $p = 4$ was utilized for the FVs in this analysis. Thus, only the first four reconstructed components were used to build the baseline FV matrix. The data measured on the healthy blade was divided into two sets, a training set and a testing one, each of which was built using 21 signals/vectors. The training set was used to create the baseline FV matrix where the newly observed FVs are to be compared.

Accordingly, the baseline FV matrix $\mathbf{T}_B$ was constructed by using $s = 21$ FVs with a dimension $p = 4$ (see Section 3.2.2). The Mahalanobis distance of each observation to the baseline matrix $\mathbf{T}_B$ was measured to determine the DI corresponding to each observation (see Section 3.2.4).

In order to visualize the methodology performance, the damage indices obtained when using the signals from sensor 4 at the TE are presented in Figure 3 when the blade was excited at the different actuator locations. It can be seen that for the four cases, the damaged cases were clearly detected. The number of false alarms (when the blade was not damaged

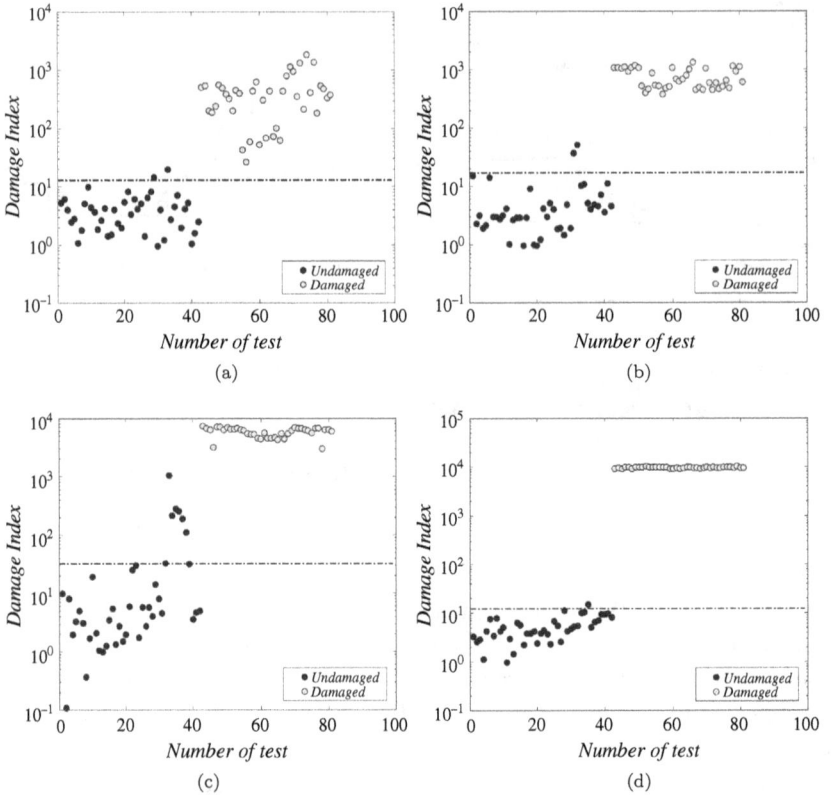

Figure 3.   Mahalanobis distance damage index (DI) for different sensors and actuators.
(a) Sensor TE 4, Actuator 1, (b) sensor TE 4, Actuator 2, (c) sensor TE 4, Actuator 3
and (d) sensor TE 4, Actuator 4.

but the system classified it as damaged) and missed damage cases was low
for most cases except for a couple of exceptions.

It can be seen from Table 1 that most of the damaged as well as
the healthy cases were correctly classified for most actuator and sensor
positions. An exception is the healthy cases classified as damaged for the
position A3 of the actuator and for a number of the sensor positions. This
actuator is inside the blade and is rather far away from the damage. Actuators 1 and 2 are closer to the damage as compared to actuators 3 and 4,
and the results for the cases of A1 and A2 are in general better than the
results for the actuator positions A3 and A4.

In terms of sensors from Figure 2, it should be appreciated that sensors 5 and 6 are the ones closest to the damage (sensor 5 is the closest

Table 1. Percentage of correct classification of healthy and damaged observations for the SSP34 m-WTB.

Actuator	(%)	Sensors TE									
		1	2	3	4	5	6	7	8	9	10
A1	Correct H	100	98	95	95	100	100	100	100	95	95
	Correct D	100	100	92	100	62	56	100	100	100	100
A2	Correct H	95	86	76	95	98	98	98	86	95	98
	Correct D	100	100	100	100	100	100	100	100	100	100
A3	Correct H	69	74	98	83	83	76	81	83	71	64
	Correct D	100	100	100	100	100	100	100	79	79	72
A4	Correct H	83	100	86	98	79	90	93	88	95	83
	Correct D	100	92	100	100	100	100	100	100	100	69

Note: Threshold at risk of false alarm probability equal to $\alpha = 0.01$ for lognormal distribution. The FV dimension is 4 (T1–T4). Total healthy observations = 42, total damaged observations = 39.

one). It can be seen from Table 1 that these sensors give the worst results in terms of identifying the damaged cases, especially for the actuation in position A1, which is also the closest one to the damage.

3.5. *A case study for rolling element fault diagnosis*

This part considers the application of the method for fault assessment described in Section 3.3.

3.5.1. *The case study*

The bearing vibration data were obtained from the test rig of Case Western Reserve University (CWRU). The data-bearing center[38] shown in Figure 4 consists of a 3 HP three-phase induction motor: a dynamometer. The drive end bearing (SKF 6025 deep grove ball bearing) data was used in this analysis. An electrical discharge machine was used to introduce single-point faults in the bearing raceways and ball elements of different bearings with fault diameters of 0.007, 0.014, and 0.021 inches and a depth of 0.011 inches. The bearing vibration data sets were obtained at a sampling rate of 12 kHz for different fault sizes and at speeds varying from 1730 rpm to 1797 rpm. The data for the ORF were taken with the fault position centered at the 6 o'clock position with respect to the load zone.

Figure 4.   The bearing test rig of CWRU.[38] (1) Induction motor, (2) accelerometer position, (3) torque transducer, and (4) dynamometer.

The results presented for demonstration are based on the signals obtained at 1730 rpm for a healthy bearing (H), a bearing with an IRF, a bearing with a ball fault (BF) and a bearing with an ORF with a fault size of 0.007 inch in diameter for all the fault types. The first part of the diagnosis process is the detection which as explained earlier, is done on the basis of a threshold for the Mahalanobis distance. In this case, the threshold is selected so that 99% of the data from the training sample are below the threshold. Figure 5 presents a raw signal corresponding to a healthy bearing.

3.5.2. *Fault assessment*

3.5.2.1. Fault detection

The decomposition into PCs will result in building the baseline space. For the purposes of visualization only the first three PCs were used, which in this case contained 80% of the total variance of the data, but even with these the results are quite impressive.

Figure 6 presents the baseline data set corresponding to the healthy bearings and the projections of the measured data for the cases of different sizes and different fault types. It can be visually perceived that the classes/sets corresponding to the different fault conditions are very well separated. All the faulty conditions are correctly identified as such — all

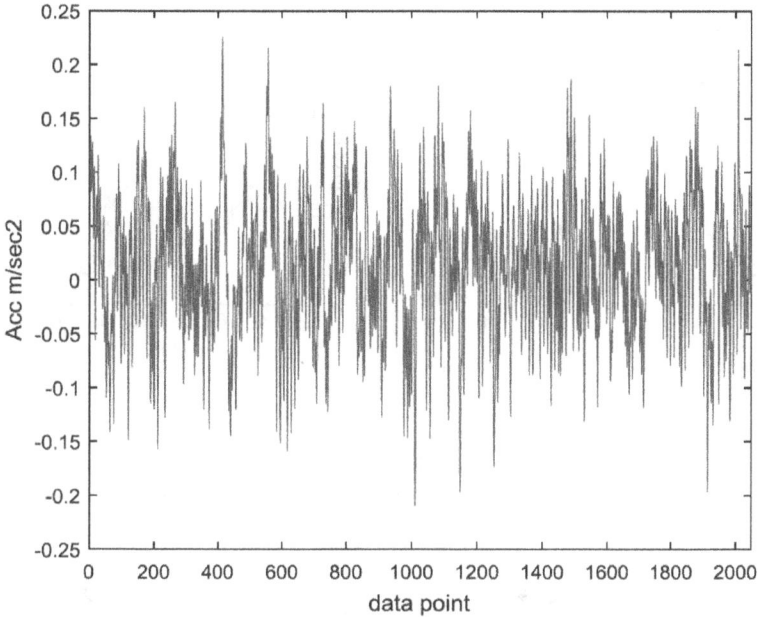

Figure 5. Raw signal from healthy bearing.

the data corresponding to the three different fault categories — IRF, ORF, and BF are quite away from the set threshold. Table 2 summarizes the detection results. The results are based on the testing sample, which in this case was built using 210 signals, 30 signals from the healthy category and 180 signals from the faulty category. It can be seen that all the faulty cases are correctly identified as faulty. Around 93.3% of the healthy cases are correctly recognized as healthy, while about 6.7% are miss classified as faulty.

3.5.2.2. Fault identification: Type and size estimation

Once the presence of a fault is established, that is the detection stage is fulfilled, the next two stages are the fault type determination and the fault size estimation. In this particular case, the data one is in possession of are signals from healthy bearings (H), bearings with an IRF, bearings with an ORF, and bearings with a BF. The task at the fault type identification stage is to distinguish between these three categories. As was explained earlier (see Section 3.3.1), this is done using training samples for each of the categories in order to build the corresponding reference spaces/matrices.

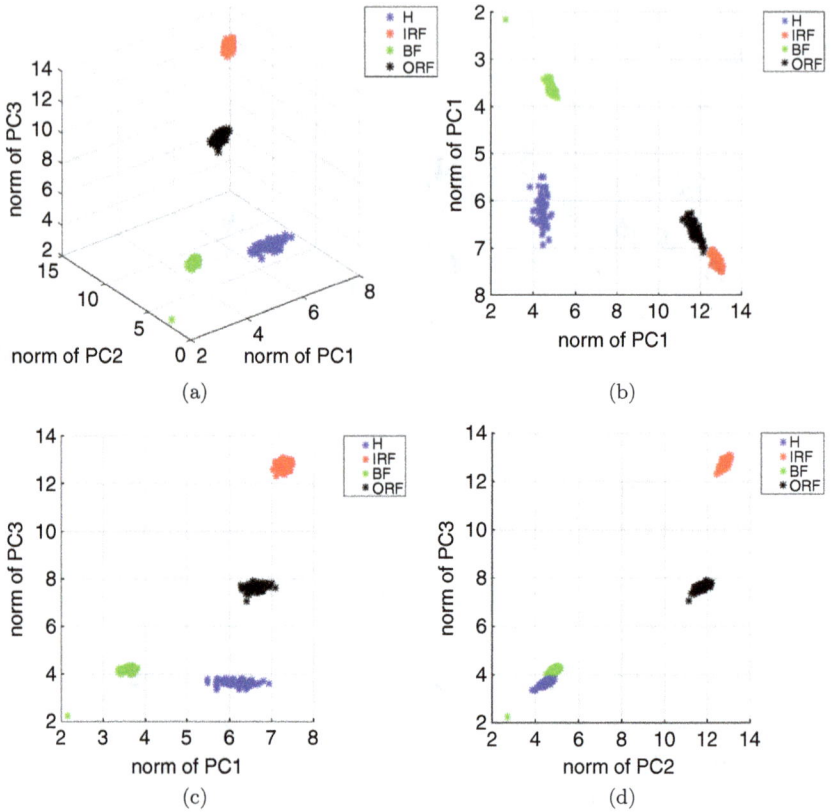

Figure 6.   Projection of the signals from different categories on the baseline space.

Table 2.   Confusion matrix for the detection of rolling element bearing faults.

Actual/identity class (%)	Healthy	Faulty
Healthy	93.3	6.7
Faulty	0	100

Then the Mahalanobis distance of a new projection to each of the categories is measured. Eventually, the signal is categorized to the set to which its distance is the smallest one. In this case, half of the measured signal segments corresponding only to small faults (30 signal segments altogether), were used to build a training sample. The rest of the signals, that is $4 \times 30 = 120$

Table 3. Percentage of correctly classified and misclassified faults for the case of 1730 rpm.

Real/identified fault type (%)	H	IRF	ORF	BF
H	100	0	0	0
IRF	0	100	0	0
ORF	0	0	100	0
BF	0	0	0	100

Table 4. Confusion matrix for size estimation/ categorization for IRF 1730 rpm.

Correct/identified fault size (%)	H	S	M	L
H	100	0	0	0
S	0	100	0	0
M	0	0	100	0
L	0	0	0	100

in this case, were used as a testing sample. For rotational speed of 1730 rpm and for the case of three features, i.e. 3 PCs were used to make the baseline space, all the signals from the testing sample were correctly classified. And the confusion matrix is given in Table 3.

It can be appreciated form the results in Table 3 that all the faults from the four different fault type categories were correctly classified to the right category.

Once the type of the fault is established, the fault size was estimated. In this case, the data available was for three different fault sizes — $0.007''$, $0.014''$, and $0.021''$. So these three sizes were assigned to three different categories namely small (S), medium (M), and large faults (L). The size identification is performed the same way as the type identification. That is, training samples were used made of 30 signals from each category. Then, each signal from the testing sample was assigned to its closest category in the sense of Mahalanobis distance. As a demonstration, the confusion matrix for the particular case of rotational speed of 1730 rpm and for the case of IRF is given in Table 4.

It can be seen that all the signals were correctly classified to the right size category.

3.6. *Discussion and conclusions*

This chapter offers a study on the application of data analysis methods for the purposes of vibration-based structural and machinery health and condition monitoring. Data-driven methods which are based on pure data analysis have a lot of potential for monitoring applications in structures and in machinery as they do not assume any model or linearity and thus offer a more general application which can be used for different structures/machinery and for a variety of faults/damages. Therefore, even though the methods considered in this study were developed and tested for particular applications, they can be easily used for different purposes of structural and machinery monitoring.

In particular, the methodology considered is based on SSA, which decomposes the measured vibration signals into new independent components which contain most of the variance of the original data. Eventually, the initial signals can be reconstructed and the precision of the decomposition using different numbers of components can be evaluated. In this study, SSA is applied in different modifications for (1) SHM and (2) for machinery condition monitoring.

In the first application, a method for detecting structural anomalies (a SHM method) is developed, which uses the reconstructed components and compares them to a reference state/space based on the reconstructed signals from the healthy/pristine condition of the structure. The method uses a threshold in order to distinguish between signals, measured on the healthy structure and signals measured in a damaged condition.

In the second application, a method for rolling element fault detection and identification is developed, which applies SSA in different ways so that a baseline space is built using the decomposed signals from the healthy/baseline condition. Consequently, a new signal embedded into an embedding matrix is projected onto the baseline space in order to evaluate its similarity to the baseline/healthy condition.

The method for SHM was developed primarily for delamination diagnosis purposes, and the method for machinery diagnosis was specifically developed for rolling element diagnosis purposes. It should be mentioned that the method for SHM has been applied to a number of applications (although just one is mentioned within this study) including for delamination detection in composite plates[39] and also for a lab-scale turbine blade.[34] It demonstrated very good results for all these applications.

The method for rolling element bearings has been applied for three case studies, two of which are experimental test rigs — one at Strathclyde University and one at the University of Torino.[40] In this study only the application using the data from CWR University is shown as it is most demonstrative in terms of data variety from different fault conditions. But it should be mentioned that the methodology for rolling element fault diagnosis gave very high percentage of correct classification (close to 100%) for all case studies. Thus, it has been validated and verified for different applications of rolling element bearing fault diagnosis. It should be mentioned that this methodology provides a rather simple approach for rolling element bearing fault diagnosis and at least at the first stage, the detection can be quite easily applied if one is in possession of data for the baseline/healthy state only.[35] The subsequent stages for fault type identification and severity estimation as currently developed require data from damaged conditions, which should be usually taken from the same machine. The severity estimation without supervision that is based on the healthy/baseline condition data is under consideration for further research.

The case study for the turbine blade application was primarily developed for the purposes of preliminary research and one of our goals was to test the best position of the application of the excitation and the sensor(s). As can be seen from the results discussed, the method achieves quite good separation between the faulty and the healthy conditions for most configurations of the sensors and the actuators. The separation is not good when both/or any — the sensor and/or the actuator — are close to the damage. The best results are obtained when both the sensor and the actuator are not very close but they are also not too far away from the damage. Our best results are for sensors 2, 3, and 4 when the damage is between sensors 5 and 6 and for actuation point 3. More research is needed in order to develop the method to an application stage.

Future research in the directions of application of SSA for the purposes of machinery and SHM will be oriented in two main areas: the development of better classification and diagnosis methods and the development of fully applied methods that can be incorporated for practical monitoring. The methods used for classification purposes as presented earlier in this chapter can be further developed in order to make the process easier and more automatic. Pattern recognition methods can be applied and e.g. linear or nonlinear boundaries between the different classes can be designed for the purposes of recognition. Also, the development of more unsupervised methods when there is no or minimal information about the damaged states for the machinery and the structural diagnosis are needed.

Acknowledgement

The authors acknowledge the help and the collaboration with Dr Dmitri Tcherniak from Brüel & Kjær S/V Measurements who did the experiment with the turbine blade and kindly provided the data.

References

1. Doebling, S. W., Farrar, C. R., Prime, M. B., and Shevitz, D. W. Damage identification and health monitoring of structural and mechanical systems from changes in their vibration characteristics: A literature review; Los Alamos National Lab., NM (United States) (1996).

2. Carden, E. P. and Fanning, P. Vibration based condition monitoring: A review. *Structural Control and Health Monitoring* **3**(4), pp. 355–377 (2004).

3. Worden, K. and Manson, G. The application of machine learning to structural health monitoring. *Philosophical Transactions of the Royal Society of London A: Mathematical, Physical and Engineering Sciences* **365**(1851), pp. 515–537 (2007).

4. Sohn, H. and Farrar, C. R. Damage diagnosis using time series analysis of vibration signals. *Smart Materials and Structures* **10**(3), p. 446 (2001).

5. Bishop, C. M. *Neural Networks for Pattern Recognition.* Oxford University Press (1995).

6. Yao, R. and Pakzad, S. N. Autoregressive statistical pattern recognition algorithms for damage detection in civil structures. *Mechanical Systems and Signal Processing* **31**, pp. 355–368 (2012).

7. Tandon, N. and Choudhury, A. A review of vibration and acoustic measurement methods for the detection of defects in rolling element bearings. *Tribology International* **32**(8), pp. 469–480 (1999).

8. Basseville, M., Mevel, L., and Goursat, M. Statistical model-based damage detection and localization: Subspace-based residuals and damage-to-noise sensitivity ratios. *Journal of Sound and Vibration* **275**(3), pp. 769–794 (2004).

9. Kopsaftopoulos, F. and Fassois, S. Vibration based health monitoring for a lightweight truss structure: Experimental assessment of several statistical time series methods. *Mechanical Systems and Signal Processing* **24**(7), pp. 1977–1997 (2010).

10. Taylor, S. J. *Modelling Financial Time Series.* World Scientific Publishing (2007).

11. Wei, W. W. S. *Time Series Analysis.* Addison-Wesley, Reading, MA (1994).
12. Hassani, H. Singular spectrum analysis: Methodology and comparison. *Journal of Data Science* **5**(2), pp. 239–257 (2007).
13. Jolliffe, I. *Principal Component Analysis.* Wiley, Online Library (2002).
14. Mujica, L. E., Rodellar, J., Fernández, A. and Güemes, A. Q-statistic and T2-statistic PCA-based measures for damage assessment in structures. *Structural Health Monitoring* 1475921710388972 (2010).
15. Johnson, M. Waveform based clustering and classification of AE transients in composite laminates using principal component analysis. *NDT & E International* **35**(3), pp. 367–376 (2002).
16. Kilundu, B., Chiementin, X., and Dehombreux, P. Singular spectrum analysis for bearing defect detection. *Journal of Vibration and Acoustics* **133**(5), p. 051007 (2011).
17. Muruganatham, B., Sanjith, M., Krishnakumar, B., and Satya Murty, S. Roller element bearing fault diagnosis using singular spectrum analysis. *Mechanical Systems and Signal Processing* **35**(1), pp. 150–166 (2013).
18. Lopez, I. and Sarigul-Klijn, N. Effects of dimensional reduction techniques on structural damage assessment under uncertainty. *Journal of Vibration and Acoustics* **133**(6), 061008 (2011).
19. Yan, A.-M., Kerschen, G., De Boe, P., and Golinval, J.-C. Structural damage diagnosis under varying environmental conditions — Part I: A linear analysis. *Mechanical Systems and Signal Processing* **19**(4), pp. 847–864 (2005).
20. Yan, A.-M., Kerschen, G., De Boe, P., and Golinval, J.-C. Structural damage diagnosis under varying environmental conditions — Part II: Local PCA for non-linear cases. *Mechanical Systems and Signal Processing* **19**(4), pp. 865–880 (2005).
21. González, A. G. and Fassois, S. A supervised vibration-based statistical methodology for damage detection under varying environmental conditions & its laboratory assessment with a scale wind turbine blade. *Journal of Sound and Vibration* **366**, pp. 484–500 (2016).
22. Ghil, M., Allen, M., Dettinger, M., Ide, K., Kondrashov, D., Mann, M., Robertson, A. W., Saunders, A., Tian, Y., and Varadi, F. Advanced spectral methods for climatic time series. *Reviews of Geophysics* **40**(1), pp. 3-1–3-41 (2002).
23. Basilevsky, A. and Hum, D. P. Karhunen-Loeve analysis of historical time series with an application to plantation births in Jamaica. *Journal of the American Statistical Association* **74**(366a), pp. 284–290 (1979).
24. Hassani, H. and Thomakos, D. A review on singular spectrum analysis for economic and financial time series. *Statistics and its Interface* **3**(3), pp. 377–397 (2010).
25. Yiou, P., Sornette, D., and Ghil, M. Data-adaptive wavelets and multi-scale singular-spectrum analysis. *Physica D: Nonlinear Phenomena* **142**(3), pp. 254–290 (2000).

26. Loh, C.-H., Tseng, M.-H., and Chao, S.-H. Structural Damage Assessment Using Output-Only Measurement: Localization and Quantification. In *ASME 2013 Conference on Smart Materials, Adaptive Structures and Intelligent Systems*, American Society of Mechanical Engineers, pp. V002T05A001–V002T05A001 (2013).

27. Chao, S.-H. and Loh, C.-H. Application of singular spectrum analysis to structural monitoring and damage diagnosis of bridges. *Structure and Infrastructure Engineering* **10**(6), pp. 708–727 (2014).

28. Salgado, D. and Alonso, F. Tool wear detection in turning operations using singular spectrum analysis. *Journal of Materials Processing Technology* **171**(3), pp. 451–458 (2006).

29. Kilundu, B., Dehombreux, P., and Chiementin, X. Tool wear monitoring by machine learning techniques and singular spectrum analysis. *Mechanical Systems and Signal Processing* **25**(1), pp. 400–415 (2011).

30. Muruganatham, B., Sanjith, M., Kumar, B. K., Murty, S., and Swaminathan, P. Inner Race Bearing Fault Detection Using Singular Spectrum Analysis. In *Communication Control and Computing Technologies (ICCCCT)*, International Conference on 2010 IEEE, pp. 573–579 (2010).

31. Garcia, D. and Trendafilova, I. A multivariate data analysis approach towards vibration analysis and vibration-based damage assessment: Application for delamination detection in a composite beam. *Journal of Sound and Vibration* **333**(25), pp. 7036–7050 (2014).

32. Zabalza, J., Ren, J., Zheng, J., Han, J., Zhao, H., Li, S., and Marshall, S. Novel two-dimensional singular spectrum analysis for effective feature extraction and data classification in hyperspectral imaging. *IEEE Transactions on Geoscience and Remote Sensing* **53**(8), pp. 4418–4433 (2015).

33. Sohn, H., Czarnecki, J. A., and Farrar, C. R. Structural health monitoring using statistical process control. *Journal of Structural Engineering* **126**(11), pp. 1356–1363 (2000).

34. García, D., Tcherniak, D., and Trendafilova, I. Damage assessment for wind turbine blades based on a multivariate statistical approach. *Journal of Physics: Conference Series* **628**(1), 012086 (2015).

35. Al-Bugharbee, H. R. S. Data-driven methodologies for bearing vibration analysis and vibration based fault diagnosis. University of Strathclyde (2016).

36. Kantz, H. and Schreiber, T. *Nonlinear Time Series Analysis*. Cambridge University Press, Vol. 7 (2004).

37. Larsen, G. C., Berring, P., Tcherniak, D., Nielsen, P. H., and Branner, K. Effect of a damage to modal parameters of a wind turbine blade. In *EWSHM-7th European Workshop on Structural Health Monitoring* (2014).

38. CWRUBDCW, The Case Western Reserve University Bearing Data Center Website (2014).

39. Palazzetti, R., Garcia, D., Trendafilova, I., Fiorini, C., and Zucchelli, A. An investigation in vibration modelling and vibration-based monitoring for

composite laminates. In *26th International Conference on Noise and Vibration Engineering* (2014).

40. Tabrizi, A. A., Al-Bugharbee, H., Trendafilova, I., and Garibaldi, L. A cointegration-based monitoring method for rolling bearings working in time-varying operational conditions. *Meccanica* **52**(4), pp. 1201–1217 (2017).

Chapter 3

Experimental Investigation of Delamination Effects on Modal Damping of a CFRP Laminate, Using a Statistical Rationalization Approach

Majid Khazaee[*], Ali Salehzadeh Nobari[*,†,‡] and M. H. Ferri Aliabadi[†]

[*]*Department of Aerospace Engineering, Amirkabir University of Technology*
Tehran 15875-4413, Iran

[†]*Department of Aeronautics Engineering, Imperial College London*
London SW7 2AZ, UK
[‡]*a.salehzadeh-nobari@imperial.ac.uk*

As an attempt towards damage detection in composites, in this chapter, an experimental investigation is presented aiming at extending the current understanding of how delamination will affect the vibration characteristics of carbon fiber-reinforced plastics (CFRP). Different percentages of delamination have been applied artificially on CFRPs, and experimental modal analysis has been performed for both natural frequency and modal loss factor parameters. It is a well-established fact that, due to the global nature of lower modes' natural frequencies, small defects have small effects on these modal parameters and hence, generally speaking, natural frequency is not a good feature for damage indication. On the other hand, modal loss factor can be considered as a good damage index (DI), especially, if the healthy structure is of low damping characteristic. However, the problem with modal loss factor is that it is hard to identify reliable values for this parameter and usually identified values have high scatter. In order to be able to identify reliable modal loss factors, the modal parameter extraction techniques with high order of accuracy in damping identification, i.e. circle fit (CF) and line fit (LF) methods, are used, leading to the rationalization process being optimized, resulting in the extended line fit method, (ELFM). Using ELFM, it has been shown that both natural frequency and modal loss factor have changed due to delamination. While, as expected, natural frequency has experienced insignificant changes, modal loss factor has proved to be a highly sensitive indicator, undergoing major changes even at initial damage stages. Modal damping mechanisms and their relationship with mode shapes have been examined. The results reveal that delamination severity can be detected using modal loss factor variations.

Keywords: Health monitoring; Chopped fiber reinforced plastic (CFRP); Dilamination; Natural frequency; Modal damping; Statistical rationalization; Experimental modal analysis; Line fit method; Damage induced nonlinearity; Optimum equivalent linear frequency response function (OELF); Damping mechanism; Composites.

1. Introduction

As a result of the trend in designing more efficient mechanical systems and structures, such structures turn to be increasingly more complicated and load sensitive. As such, condition and health monitoring are becoming vital tools in various industries, especially those which are safety critical. On the material side, the use of woven fiber-reinforced composite is becoming a widespread trend due to its high specific stiffness and strength. In addition, its viscoelastic behavior renders it suitable for applications in automotive, aerospace, and marine industries.[1,2] However, it is a well-known fact that composites, in general, are vulnerable to impact, leading to delamination at impact location even for low-energy impacts. In turn, delamination can significantly deteriorate the mechanical properties of composite materials leading to catastrophic consequences. Hence, the ability to detect the type, severity, and location of damage and probable effects on the overall behavior of structures are highly important,[3] especially for safety-critical applications such as aerospace industries.

Although non-destructive testing (NDT) application for damage detection in composites is relatively recent, methods and algorithms in this field are developing rapidly. Recent methods can be broadly divided into two categories, namely: (1) methods based on medium-range frequencies, and, (2) those based on the ultra-high frequency waves known as Lamb's waves. The methods of the first category are mainly based on the modal and frequency domain data analyses performed on the measured vibration signatures of healthy and damaged structures (data-based methods) and, on some occasions, accompanied by an finite element (FE) model of structure (model-based methods). In this respect, associated works describing damage detection boil down into three main categories, based on the parameter used as damage-sensitive feature, namely; methods using natural frequency variations,[4–7] methods based on damping or energy variations,[8–14] and methods using mode shape or its curvature variations.[15,16] Using experimental techniques, Kessler studied modal parameters' variation for graphite/epoxy laminate containing damage modes.[7] The results show that frequency domain methods can be reliably used even for detection of small damages. It is a well-known fact that, since natural frequency is a global property of structures, it will not be significantly affected by small damages,[8] unless very high-order natural frequencies are visited. On the other hand, modal damping would be a much better indicator of damage, as the variation of energy dissipation due to damage is higher than stiffness variation, especially, for lightly damped structures. Also, even small damages can lead to nonlinear damping mechanisms such as friction. Hence, modal damping can be more sensitive to delamination in comparison to low-order natural frequencies.[8]

Natural frequency-based damage detection examines the effect of damage on the structural properties and, hence, on the natural frequencies of the structure. Cawley and Adams[5] suggested a technique for damage detection, localization, and quantification. Damage localization was determined by changes in measured natural frequencies and their correlation with those derived from a finite element model. Tracy investigated the variation of natural frequencies for carbon fiber laminates subjected to impact damage. The results show that mid-span delamination degraded the even modes much more than the odd modes. Moreover, for the first four modes, natural frequencies with 1/3 length delamination had a maximum of 20% effect.[17]

Kiral *et al.* experimentally studied beam vibrations using non-touching proxy sensors for different damage locations.[11] Their research indicated that damping is much more sensitive to damage than natural frequency. They showed that damping increases as the level of impact energy is increased, the increase being highly dependent on damage location. Montalvão has presented a method for experimental detection of impact-induced damage location in composite laminates, based on inverse frequency response function (FRF) in which damping coefficient was considered as the distinguishing feature.[12] Keye has developed a method to link damage location to damping variation in CFRP for aircraft panels.[10]

Although damping is more sensitive to damage than natural frequency, its application for damage identification is hindered by the fact that damping coefficients are difficult to extract with high accuracy and reliability. Hence, the basic and significant step for damage detection in composites is to understand damping nature and devise an accurate method for modal damping extraction.

Hence, the purpose of this chapter is twofold: first, to present a relatively more accurate, systematic, approach for modal damping rationalization and, second, to use the proposed method to determine the variation in modal parameters due to delamination and effect of delamination size and, eventually, relation of modal loss factor and damping mechanism and its relation to mode shapes.

2. Experimental Setup and Test Specimens

Modal testing was performed on woven symmetric $[0/90]_{4s}$ CFRP composite. Quasi-isotropic laminates were fabricated from eight-plies with dimensions of 200 mm × 200 mm × 12 mm. Each layer is a woven prepreg T300 integrated in epoxy resin ML-506 and epoxy hardener HA-11. Pristine specimens were fabricated without any imperfections. Composite laminates with artificially induced delamination of different sizes were fabricated and tested. The delamination was artificially introduced by placing vacuum bag with 5%, 10%, and 20% of the laminate area between layer 4 and 5 at the center of laminate. As will be shown later in Section 5.3, this particular zone was selected for the introduction of delaminations, in order to be able to distinguish between the effects of various damping mechanisms. Figure 1 shows the pristine and delaminated laminates before and after resin injection. In order to assure that random fabrication anomalies are minimized and accounted for, all specimens, pristine or delaminated, are fabricated under strict control and three specimens are fabricated for pristine specimen in order to test the repeatability.

The specimens were tested in a free–free configuration using nylon strings. A single-point excitation using modal hammer (Brüel & Kjær Impact hammer Type 8202) was performed and a force transducer (Brüel & Kjær Type 8200) captured the impulse force. Figure 2 shows the test setup and free-free boundary condition. The hammer tip and the impulse energy were controlled, in order to prevent creation of further damages.

Carbon laminate responses were measured in 0–800 Hz frequency range (covering five bending modes) with a dual-channel analyzer B&K Type 2035. To provide a smooth mode shape during modal analysis, FRFs were recorded at a 25-points network shown in Figure 3. A 2 [gr] accelerometer was located at the laminate center, and consistency of measurement was checked to make sure that the accelerometer weight is not affecting the results.

3. Proposed Modal Analysis Approach: Extended Line Fit Method (ELFM)

This research relies on modal parameters extracted from experimental modal analysis for pristine and damaged laminates. Experimental modal analysis deals with extraction of modal parameters from measured FRF data; therefore, it is a path from response to modal model. The challenging question is the accuracy of modal parameters derived from measured data. It is challenging because the answer depends on the methods and models used in the modal analyses process. In general, it is believed that single

(a)

(b)

Figure 1. Pristine and damaged CFRP specimens (a) before and (b) after resin injection.

degree-of-freedom, single FRF modal analysis techniques, single degree-of-freedom technique (SDFT) are more accurate and more informative,[18] especially, when modal damping is of interest. Also, SDFTs give the analyst more control over rationalizing of the analyses results. The main shortcoming of SDFT is that it is time consuming. Here, it has been endeavored to use the advantages of SDFT and remove its shortcoming by developing an automated, statistical, rationalizing process.

In this respect, LF method due to its accuracy[19] and practical application in SDFT is used in conjunction with some statistical rationalization analysis and the combination is called as ELFM. Also, results from Extended Circle Fit Method (ECFM) are presented for comparison. ECFM

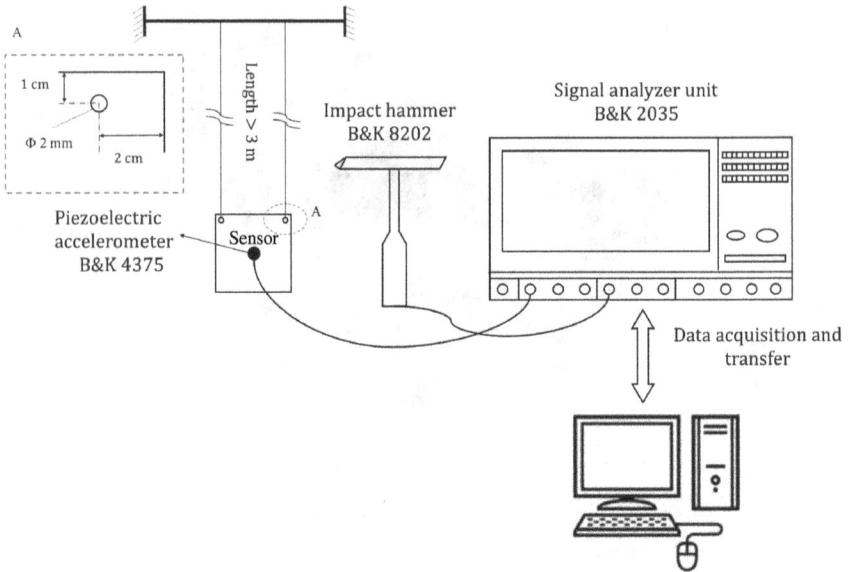

Figure 2.   Test setup for modal experiments.

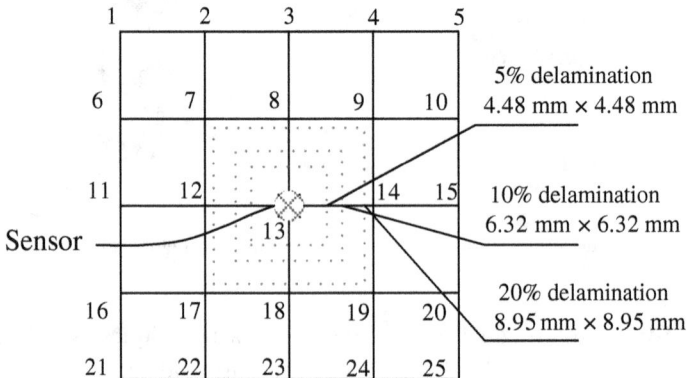

Figure 3.   Laminate mesh vertexes for modal analysis, sensor location, and delamination areas.

has the same procedure as ELFM except that circle fitting is replaced with the line fitting. Figure 4 shows the algorithm for ELFM, which has three main stages.

Stage I deals with the excitation and extraction of FRF for acceleration, Accelerance. In stage I, all the FRFs from single input single output experiments are recorded.

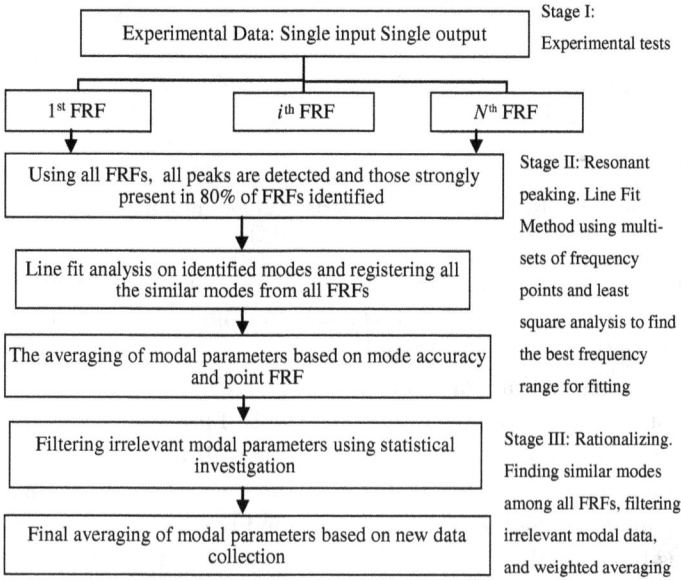

Figure 4.   Three stages of ELFM algorithm.

Stage II deals with finding the modes with strong presence in 80% of FRFs and performing modal analysis on such modes of each FRF individually.

Receptance of a system with structural damping model and for a small frequency range $[\omega_L \omega_U]$ near the rth mode can be expressed as[20]:

$$x_i/F_k = \alpha_{ij}(\omega) \approx {}_r(A + jB)_{ik}/(\omega_r^2 - \omega^2 + \omega_r^2 \eta_r j) + \text{residual} \quad (1)$$

where x_i is the displacement response at the ith degree of freedom; F_k is the excitation force (here from a modal hammer) at k^{th} degree of freedom; η_r is the loss factor of the rth mode; ω_r is the undamped natural frequency of the rth mode; and $A + jB$ is the complex modal constant of the rth mode. In CF method, the residual constant is ignored while in LF method it is taken into account. In order to eliminate the effect of residual, the value of α_{ij} at a fixed point $\omega_L < \Omega < \omega_U$ is subtracted from Equation (1). After some manipulations,[20] it yields:

$$\Delta = \frac{\omega^2 - \Omega^2}{\alpha - \alpha_\Omega} = [m_R \omega^2 + c_R] + j[m_I \omega^2 + c_I] = \text{Re}(\Delta) + j\text{Im}(\Delta) \quad (2)$$

As Equation (2) shows, real and imaginary parts of Δ are straight lines with slopes m_R and m_I with y-axis distances c_R and c_I, respectively. If different

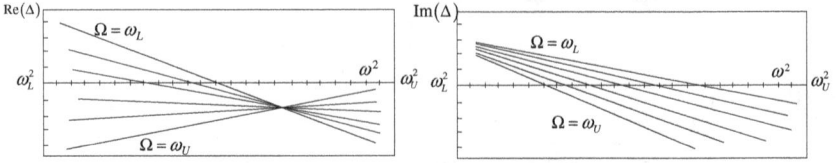

Figure 5.   Real and imaginary parts of Δ for different fix points.

fix points are selected within the band close to resonant frequency, a family of linear curves will be obtained as shown in Figure 5. For further details about LF method, see Refs. 19 and 20.

The LF method has some advantages over CF method,[19] which makes it the best choice for modal analysis in stage II. First, curve fitting for lines is easier in comparison to complicated CF, which makes it easier to spot noisy data from expected forms, such as experimental noises, inappropriate damping model, or nonlinearities. The second reason behind using inverse functions-based method is that the most significant data are those far from resonant, and since data around resonant are generally the most difficult points to record, LF method employs this advantage.

In the final stage, stage III, all modal data from multi-frequency responses are gathered to form a matrix containing all vibration modes, then filtering and weighted averaging is performed. For this purpose, first similar modes from all the FRF are searched, based on the similarity of natural frequencies. Then, a matrix containing all the natural frequencies from all FRFs is obtained:

$$
[\omega]_{\text{total}} =
\begin{array}{c c c c c c c}
\text{FRF1} & \text{FRF2} & \cdot & \text{FRF}m & \cdot & \cdot & \text{FRF}N \\
\left[\begin{array}{ccccccc}
\omega_{11} & \omega_{21} & \cdot & \omega_{m1} & \cdot & & 0 \\
0 & 0 & \cdot & \cdot & \cdot & & \omega_{N1} \\
\omega_{1j} & \cdot & \cdot & 0 & \cdot & & \cdot \\
\cdot & \omega_{2k} & \cdot & \omega_{mk} & \cdot & & \cdot \\
\cdot & \cdot & \cdot & \cdot & \cdot & & 0 \\
0 & \omega_{2l} & \cdot & \omega_{ml} & \cdot & & \omega_{Nl}
\end{array}\right]
\end{array}
\tag{3}
$$

where $[\omega]_{\text{total}}$ is the frequency matrix containing all of the natural frequencies from all experimental FRFs. In $[\omega_{ij}]_{\text{total}}$, subscript i shows the FRF number, subscript j shows the mode number. So for $[\omega]_{\text{total}}$, each column represents natural frequencies from one frequency response. Hence, zero in $[\omega]_{\text{total}}$ matrix indicates the absence of a mode in an FRF.

Weighted averaging is required in the next step in stage III, based on the fact that the accuracy of the same extracted modal parameters is different for different FRFs $[\bar{\omega}]_{\text{total}}$ can be calculated as:

$$[\bar{\omega}]_{\text{total}} = \begin{bmatrix} \dfrac{1}{\sum_{i=1}^{N} W_{1i}} \sum_{i=1}^{N} W_{1i} [\omega_{i1}]_{\text{total}} & \omega_{i1} \neq 0 \\[4ex] \dfrac{1}{\sum_{i=1}^{N} W_{2i}} \sum_{i=1}^{N} W_{2i} [\omega_{i2}]_{\text{total}} & \omega_{i2} \neq 0 \\[2ex] \vdots \\[2ex] \dfrac{1}{\sum_{i=1}^{N} W_{Li}} \sum_{i=1}^{N} W_{Li} [\omega_{iL}]_{\text{total}} & \omega_{iL} \neq 0 \end{bmatrix} \tag{4}$$

where W_{Li} is the weight of Lth mode at i^{th} FRF.

The definition of weighting parameter for each mode is derived from two factors, namely: (1) mode's strength in various FRFs identified by magnitude of its modal constant and, (2) whether FRF is point FRF or not. The higher the modal constant magnitude, the better signal to noise ratio can be expected.

Modal parameters derived from point FRFs are given higher weight in statistical analyses and averaging. The reason for this higher weight is that modal loss factor has its most accurate estimation derived from the point FRF. This can be explained as follows. From the theory of optimum linear frequency response function (OELF) it can be shown that the OELF for a nonlinear structure can be derived as[21]

$$H(\omega) = \frac{S_{xf}}{S_{ff}} \tag{5}$$

where, in Equation (5), $x(t)$ is the random response of the nonlinear system to the random excitation $f(t)$. As with the linear system FRF, OELF in Equation (5) can be used for estimation of the modal damping of the nonlinear system. However, this estimation is only exact when H in Equation (5) is a point FRF.[18] Since, for a damaged composite, weakly nonlinear behavior can kick in, hence using measured point OELF gives accurate results for modal damping which justifies higher weights for the modal parameters associated with point FRFs.

Filtering based on statistical analysis is the next step. Poor quality of data for a mode in one FRF will influence the final, rationalized modal parameter and hence it is logical to filter out irrelevant data. Here, filtering

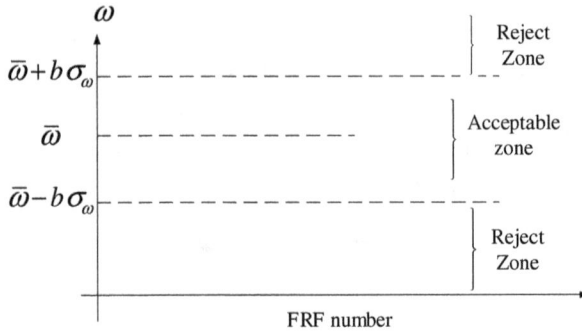

Figure 6.　Filtering of modal parameters based on the CL analyses.

is implemented based on the data Confidence Limit (CL), which is the ratio of standard deviation of a modal parameter of a particular mode from all frequency responses to their mean value. The smaller the CL for modal parameters of a particular mode, the more consistent, and hence more accurate, these modal parameters are. Figure 6 shows accept–reject zones for modal parameters, which is implemented based upon CL analysis. Equation (6) shows definition of the accept–reject band control parameter, b, based on the scatter of modal parameters of a particular mode derived from various FRFs.

$$b = \begin{cases} 1 & \sigma_{\omega_i}/\bar{\omega}_i < 5\% \\ 2 & 5\% < \sigma_{\omega_i}/\bar{\omega}_i < 15\% \\ 3 & 15\% < \sigma_{\omega_i}/\bar{\omega}_i < 25\% \\ 0 & 25\% < \sigma_{\omega_i}/\bar{\omega}_i \end{cases} \tag{6}$$

To wrap up, the ELFM has advantages associated with single degree of freedom modal analysis, namely: high accuracy for curve fitting and lower computational cost. Besides, using ELFM all the irrelevant modal data, which can be due to measurement errors or low signal to noise ratio, can be filtered out.

4. Experimental Results

4.1. Pristine carbon laminates and reliability assurance discussion

Based on the aforementioned test setup and using ELFM, FRFs and extracted modal parameters are identified for the three identical undamaged

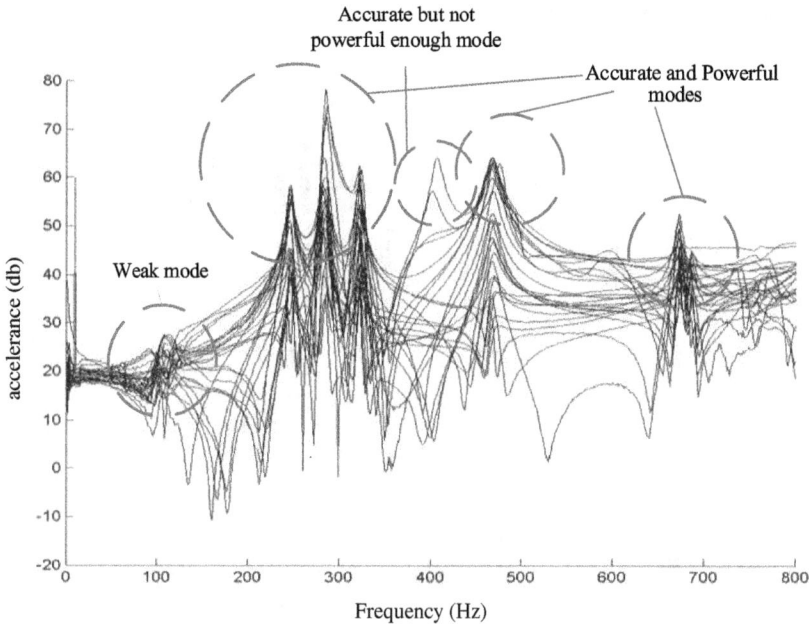

Figure 7. FRFs for all 25 excitation points in undamaged laminate, sample 1 (Weak mode: The number containing FRFs are low, Accurate: FRF is accurate around the vibration mode and powerful: The number containing FRFs are high).

laminates. As was mentioned, Accelerance FRFs for 25 excitation points have been measured. By plotting all FRFs in one diagram, Figure 7, it is possible to observe how strong presence of various modes are. If a mode has strong presence in all the FRFs, it is a strong, reliable mode.

Inspecting Figure 7, there are some discernible vibration modes in most of FRFs while some modes, such as mode in 400 Hz, cannot be seen in most FRFs. These modes are considered as strong. Five vibration modes which are accurate in modal calculation and strong in FRFs can be seen in Figure 7.

In order to investigate compatibility between tests and repeatability of measured data for three identical laminates, their related point FRFs are shown in Figure 8. Based on the point FRFs shown in Figure 8, three modes are visible for three undamaged laminates. Moreover, the FRFs have reasonable resemblance, so it can be concluded that undamaged laminates have been fabricated in a controlled way and the modal tests for them are reliable and hence have reasonably deterministic behaviors.

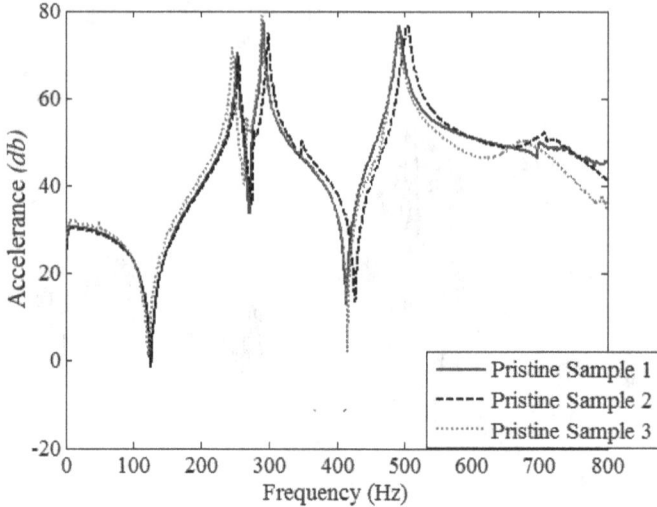

Figure 8. Point FRFs (excitation and acceleration points at 13th point) for three undamaged laminates.

Table 1. Modal parameters for three undamaged laminates based on ELFM.

Mode number	Normal sample 1 (N1)		Normal sample 2 (N2)		Normal sample 3 (N3)	
	ω_n (Hz)	ζ (%)	ω_n (Hz)	ζ (%)	ω_n (Hz)	ζ (%)
Mode I	253.10	1.10	255.05	1.21	246.11	1.09
Mode II	290.70	0.83	297.91	0.97	289.12	0.76
Weak Mode	—	—	347.44	1.12	337.66	0.72
Mode III	491.88	1.42	504.83	1.55	489.96	1.45
Mode IV	698.19	0.54	709.20	0.53	686.14	0.53

Next, ELFM is applied to pristine laminates FRFs. Table 1 shows extracted modal parameters. In all three pristine samples, four strong modes, called modes I, II, III and IV, are identified with high accuracy. It should be noted that the numbering of the modes is not in the order of their appearance in FRF. However, a weak, and hence inaccurate, mode is present in the vicinity of 340 Hz. This mode was not detected in pristine sample 1 but identified in samples 2 and 3 with low measurement accuracy. Based on Table 1, natural frequencies from 3 pristine samples are in good agreement. However, as expected, scattering for damping ratio is higher than natural frequency.

Mode shapes from experimental tests and corresponding finite element mode (FEM) shapes are shown in Figure 9. Pristine CFRP is just

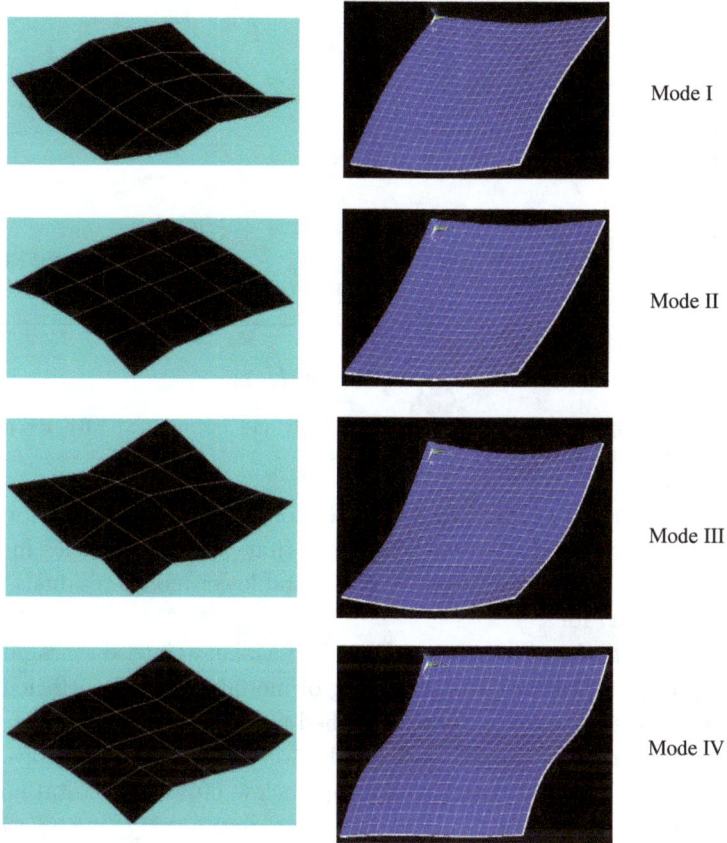

Figure 9.   Mode shapes for the four accurate and strong modes of carbon laminates.

modeled with FEM as a simple model in order to show the similarity between FEM mode shapes and their experimental counterparts. Pristine CFRP is simulated with eight perfectly bonded plies with the following material characteristics; Elastic modulus in x, y, and z directions 70 GPa, poison ratios in all directions 0.1 and shear elastic modulus in all direction 5 GPa. A mesh with 600 elements of 8-noded solid element (ANSYS SOLID185 element) is used for laminate modeling. A noticeable resemblance can be observed between identified experimental and FEM shapes.

In order to assess the quality of identified modal parameters, experimental point FRF is plotted and compared with regenerated point FRF in Figure 10. As Figure 10 indicates, in the vicinity of natural frequencies, experimental and regenerated FRFs match exactly, and hence the quality

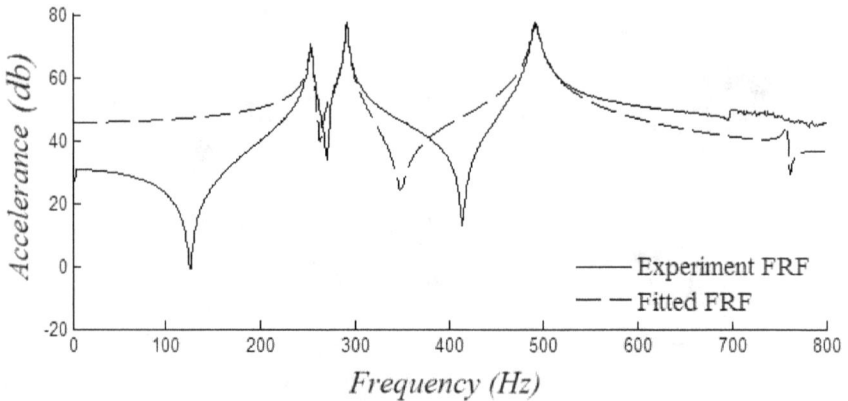

Figure 10. Comparison between experiment FRF and regenerated FRF for pristine laminate sample 1 based on ELFM results.

of extracted modal parameters and, therefore, the method used for their derivation are verified. It should be noted that the discrepancies in other frequencies are due to the effects of higher and lower modes residuals which are unaccounted for.

To illustrate the effect of stage III in EFLM, i.e. weighted averaging and filtering, Figure 11 shows the scattering of modal damping coefficients for pristine sample 2 laminate. In all identified modes, some data from certain FRFs, which are off the acceptance bands, are filtered. Filtering stops when the number of retained data points drops below 60% of the total number of data points.

The averaged results between different pair of pristine samples are shown in Table 2 with the CLs calculated for modal loss factors. Table 2 illustrates that the best consistency exists between pristine sample N1 and N3, since this pair has the least CL values. Hence, the averaged results between N1 and N3 are selected as the final modal parameters for pristine samples.

4.2. Modal parameters for damaged laminates

Here, the FRFs and modal parameters for the laminates with different delamination percentages embedded in them are presented. The characteristics of delamination are presented in the test setup section. As the first attempt for monitoring changes, point FRFs for undamaged and

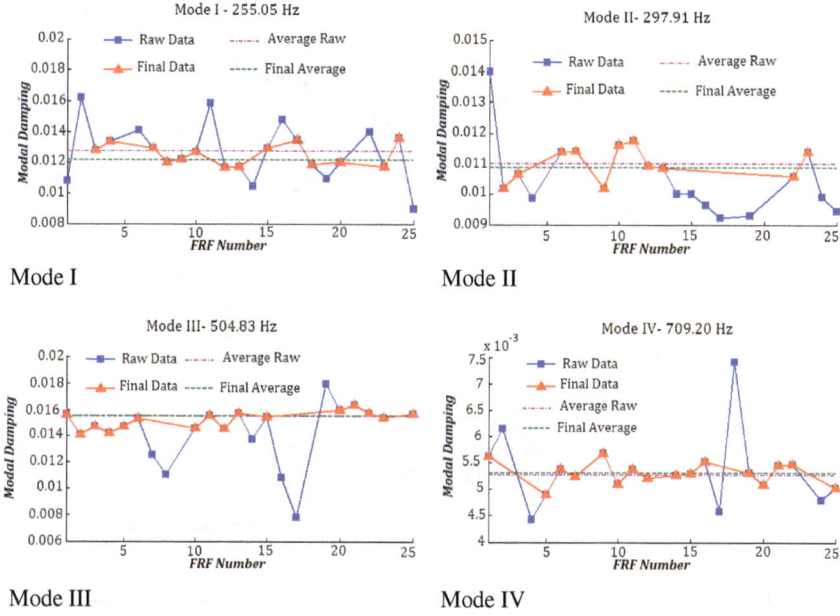

Mode I

Mode II

Mode III

Mode IV

Figure 11. The scatter of modal loss factors for pristine laminate sample 2.

Table 2. Averages and CLs for damping between different pair of pristine samples using ELFM.

	Average N1 and N2			Average N1 and N3			Average N2 and N3		
	ω_n (Hz)	ζ (%)	CL	ω_n (Hz)	ζ (%)	CL	ω_n (Hz)	ζ (%)	CL
Mode I	254.06	1.16	6.78	249.59	1.09	0.83	250.58	1.15	7.61
Mode II	294.30	0.96	18.64	289.91	0.80	6.71	293.51	0.92	25.20
Mode III	498.38	1.48	6.34	490.94	1.43	1.55	497.40	1.50	4.79
Mode IV	703.70	0.53	1.10	692.17	0.53	1.80	697.67	0.53	0.70

damaged laminates are compared in Figure 12. Due to delamination, the FRFs are changed both in resonance peaks and slopes around peaks, which means that natural frequencies and damping ratios are affected by delamination. Preliminary observations reveal that, for damaged samples, the FRF magnitudes in mode III are considerably dropped. Also, it is observed that, while mode I shows softening tendency, modes II and III show hardening behavior.

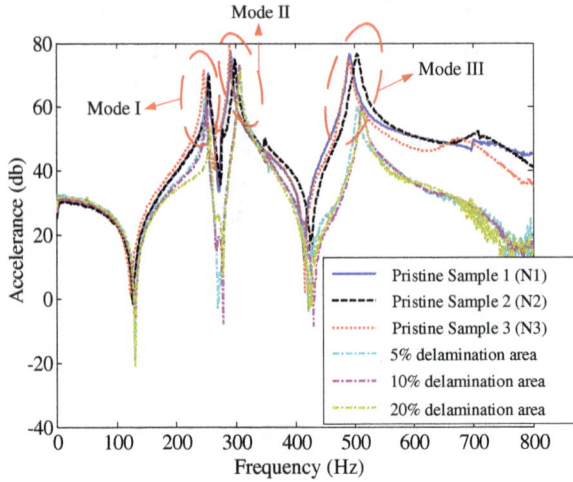

Figure 12.   Comparison between point FRFs for pristine and damaged laminates.

Table 3.   Modal parameters for pristine and delaminated laminates with different damage percentages.

	Mode I	Mode II	Mode III	Mode IV
Pristine laminates				
ω_n (Hz)	251.40	292.67	495.60	697.87
ζ (%)	1.14	0.84	1.45	0.52
CL_P (%)	0.86	1.34	1.12	1.83
5% Delamination laminate				
ω_n (Hz)	247.69	305.25	504.10	669.36
ζ (%)	0.77	0.88	1.25	0.48
CL_{D5} (%)	0.49	2.33	0.46	11.80
10% Delamination laminate				
ω_n (Hz)	250.31	305.11	512.16	691.33
ζ (%)	0.83	0.73	1.39	0.64
CL_{D10} (%)	0.99	2.43	0.64	1.53
20% Delamination laminate				
ω_n (Hz)	256.21	306.00	513.58	690.70
ζ (%)	0.71	0.83	1.39	0.62
CL_{D20} (%)	0.49	4.59	0.65	1.75

Table 3 shows modal parameters for pristine and damaged laminates with different damage percentages. Each damaged laminate has been tested five times. Among these five, the best three result sets are selected to form the final results in Table 3. For each specific mode in each CFRP specimen, modal loss factor is calculated using ELFM for 25 measured FRFs. As for the pristine samples, the concept of CL defined earlier is used for damaged laminates data filtering. CL_P, CL_{D5}, CL_{D10}, and CL_{D20} are CL for pristine, 5%, 10% and 20 % delamination area, respectively.

The sensitivity of modal parameters to delamination is different for each mode and hence, is mode-shape dependent. Hence, to predict the effect of damage on modal parameters, mode shape should also be considered.

5. Discussion on Delamination Effects on Modal Parameters

5.1. *Discussion on natural frequency variation*

Figure 13 demonstrates modal estimations for the identical undamaged laminates and for laminates with 5%, 10%, and 20% embedded delamination. Here, the variations of natural frequencies for the first four identified modes based on both ELFM and ECLM are presented. A good correlation between natural frequencies' extracted by the two methods is evident. Overall, natural frequencies' variations lie between −5% and +5% for all modes demonstrating that for low frequency modes, delamination has

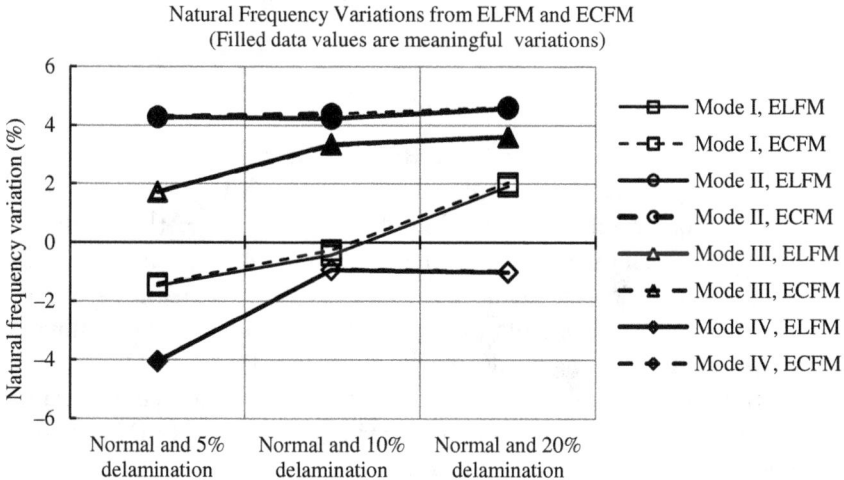

Figure 13. Comparison between natural frequencies of four modes for normal and delaminated laminates using ELFM and ECFM.

a small effect on natural frequencies. It should be noted that only filled
markers demonstrate meaningful variations, i.e. variations which are larger
than the summation of the standard deviations of pristine and damaged
specimens for the corresponding modal parameter. In other words, if σ_{rp}
and σ_{rd} are standard deviations in calculating modal parameters ω_r and
ω_{rd}, where r is the mode number and p and d stand for pristine and damaged specimens, respectively, then the variation $(\omega_r - \omega_{rd})$ is deemed as
meaningful, if:

$$|\omega_r - \omega_{rd}| > \sigma_{rp} + \sigma_{rd} \qquad (7)$$

Natural frequency variations due to delamination are less than 5% for
all four modes. While natural frequencies have been decreased in some
modes, they are increased in some other modes. This result agrees with
Tracy and Pardoen[6] when it is indicated that variation at lower modes
can be either increasing or decreasing. Based on the results from Tracy
and Pardoen[17] delamination does not always degrade the stiffness, but for
composites subjected to pure bending, delamination has little or no effect
on the stiffness. Furthermore, Cawley and Adams[5] reported that reduction in natural frequencies for single/double cuts and crushing damages
is less than 2% for maximum 3.5% damage percentage. Small changes in
natural frequencies in low frequency modes are due to the global nature
of the natural frequencies. It is also noticeable that only mode II can offer
meaningful variations for all 3 levels of damages which can be attributed
to the mode shape of this mode.

5.2. *Discussion on damping variation*

In spite of some promising reports in Refs. 4–6, as was shown, the sensitivity of natural frequencies with respect to delamination is generally low,[10]
particularly for internal delamination of the type used in this study, i.e.
delamination in mid-plane of a symmetric layup. On the other hand, the
effect of delamination on modal loss factors appears to be more promising. The variations of modal loss factors between pristine and delaminated
laminates for four strong modes are displayed in Figure 14. In order to
demonstrate the superiority of ELFM in producing more accurate results,
modal loss factors extracted using ICATS[22] multi-FRF, multi-DOF, Global
Method is also presented in Figure 14.

As is evident, unlike natural frequency, modal damping can vary up
to 30% for 5% delamination percentage. Modes I and IV have distinctive

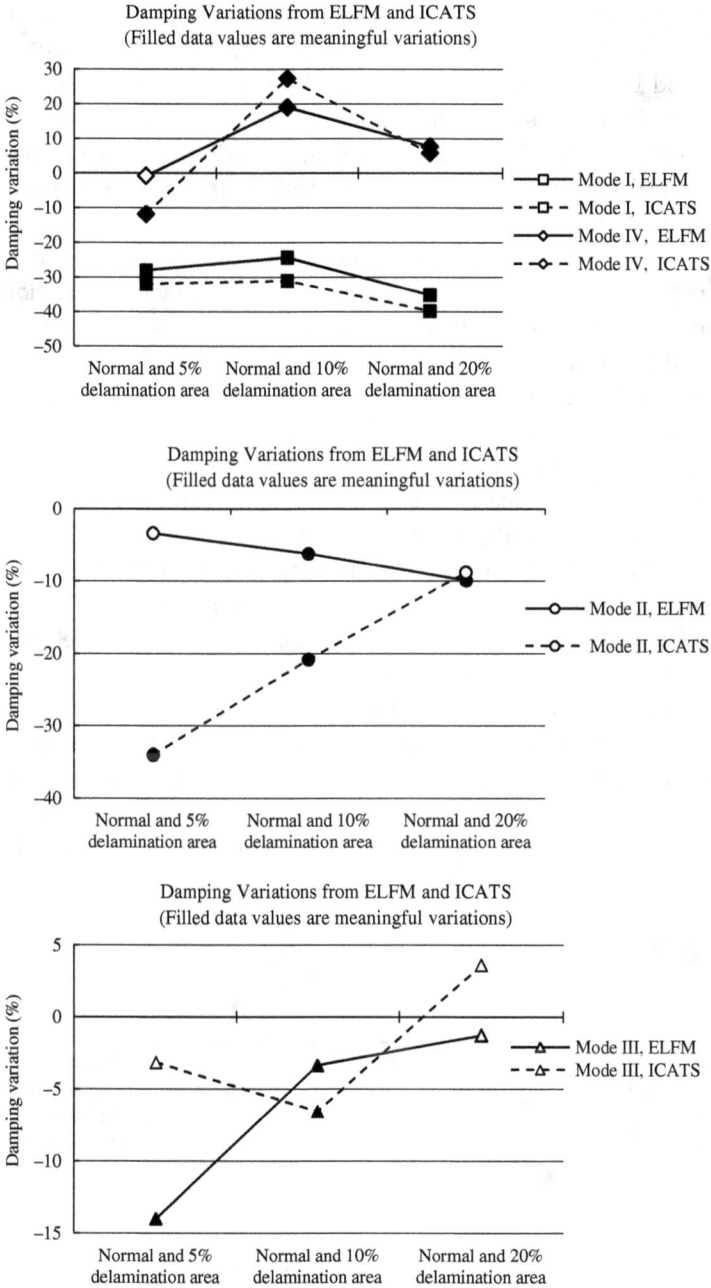

Figure 14. Comparison between modal damping in four modal modes for pristine and delaminated laminates based on ELFM and ICATS.

identical patterns for damping variations based on both modal extraction methods even though a minor difference is visible between outcome of ELFM and ICATS. On the other hand, discrepancies are evident between ELFM and ICATS results of modal loss factor variations for modes II and III. However, as ELFM lends itself to a logical rationalization scheme based on statistical filtering and weighted averaging, final results are presented based on ELFM.

Due to complicated mechanisms involved in damping, it is not straight forward to assess the reasons behind modal loss factor variations and the way they wary, so an in depth analysis is needed to distinguish the influential aspects.

Based on ELFM, modal loss factor for laminates with embedded delamination of 5%, have been decreased for all modes. However, as delamination grows, various patterns from different mode shapes are emerged. Tracing damping variations indicate that modal loss factor does not follow a regular pattern but its variation with delamination size is meaningful. If just above threshold, meaningful, variations are taken into consideration, as shown in Figure 15, it can be seen that delamination growth from 5% to 10% increases modal loss factors, while growth, from 10% to 20% delamination, makes modal loss factors to drop significantly.

Closer inspection indicates correlations between modal loss factor variations trend and their corresponding mode shapes. All modes with mode shapes with zero or small slopes in the delamination zone, i.e. modes I, II, and III, experienced a drop in modal loss factors with delamination, in comparison to pristine laminates. One the other hand, mode IV with

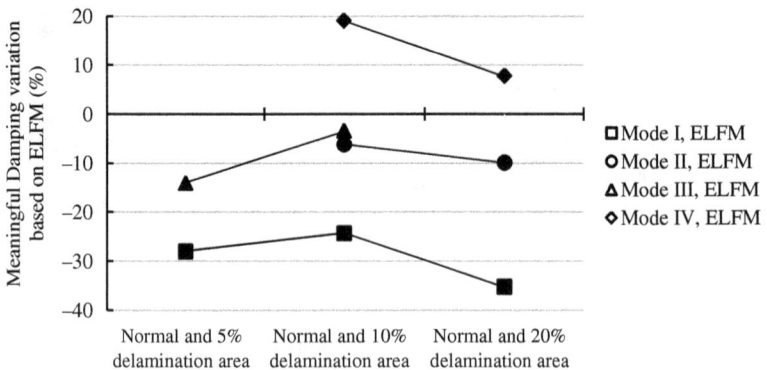

Figure 15.   Meaningful damping variations based on ELFM for four bending modes.

non-zero slope in delamination zone, show a modal loss factor increase due to delamination which can be attributed to shear effect and consequently to friction.

5.3. *Damping mechanisms in delaminated composite laminates*

Many researchers have conducted experimental studies in order to examine damping behavior variation with respect to delamination growth.[13,14,23,24] Compared to metals, composites show a higher damping capacity due to many damping mechanisms involved, such as the viscoelasticity of matrix or hysteresis mechanism in filaments.[25] Damping in composite materials is considered here as any phenomenon inside the material in which energy is dissipated. This contains, but is not limited to, (1) internal friction within each of the components material,[2,26] (2) interfacial slip at the fiber-matrix interfaces,[2,26,27] viscoplastic damping[2] and thermoelastic damping.[2]

Composite laminates consist of elastic plies embedded with viscoelastic material, i.e. resin. Matrix is the major contributor in composite damping due to viscoelasticity, although composite material exhibited elastic behavior in strongly bonded fibers to matrix.[27] Figures 16 and 17 demonstrate the point FRFs and LF analyses results for mode I and mode IV, respectively, for pristine and damaged specimens.

As is evident from Figure 16, besides the fact that delamination has decreased modal loss factor significantly, increase in delamination has two distinctive effects, (1) reduction in modal constant and (2) increase in the phase angle and hence more complex modes. The same observations can be made from Figure 17 except for the fact that, for mode IV, delamination has increased modal loss factor. This is true except for the case of 5% delamination for which loss factor is decreased slightly, an interesting observation.

These observations in Figures 16 and 17 can be attributed to various damping mechanisms effects. Regarding the nature of damping changes due to damage, some researchers qualitatively stated the main damping mechanisms.[2,3,23,24]

Saravanos and Hopking implied that viscoelastic laminate damping and interfacial friction damping appear to cause damping variations as the result of delamination.[23] Also, Chandra *et al.*[2] stated that damping due to damage is of two types, namely (1) frictional damping due to slip

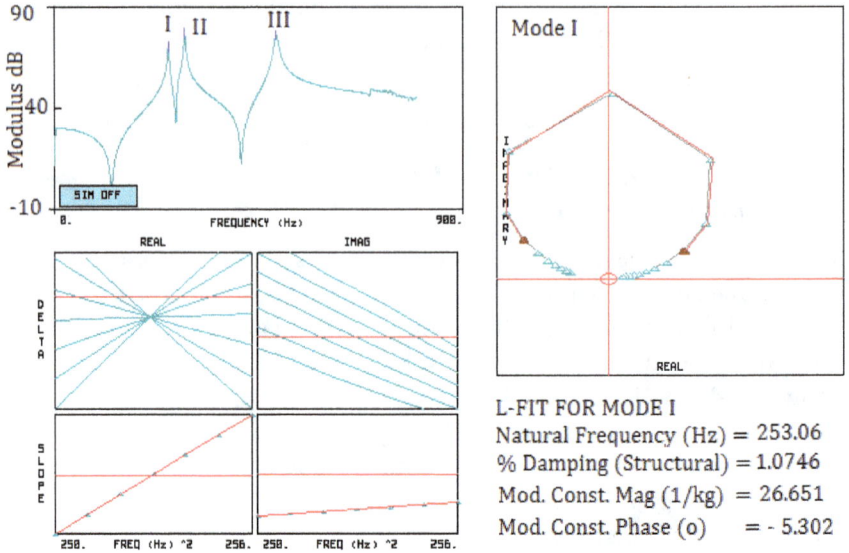

(a)

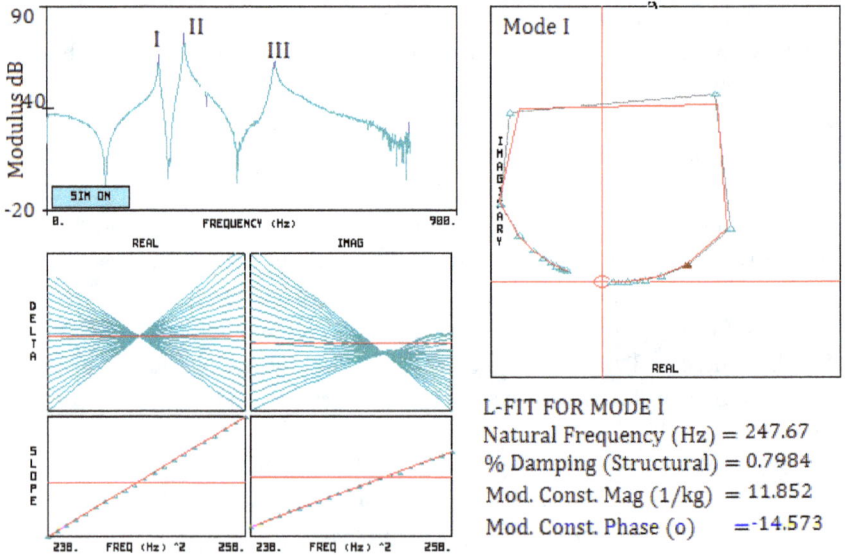

(b)

Figure 16.   Point FRFs and Mode I LF analyses for (a) pristine, (b) with 5%, (c) with 10%, and (d) with 20% delamination, respectively.

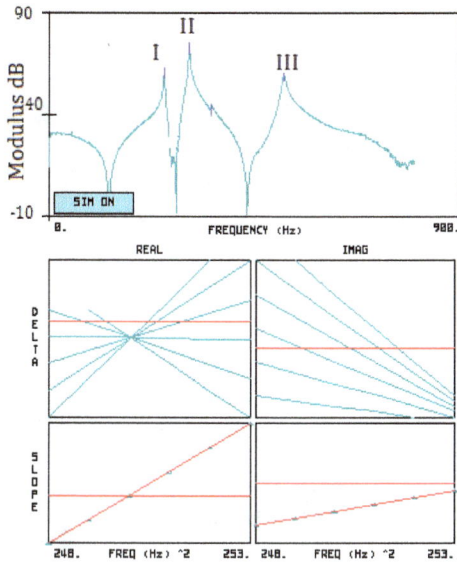

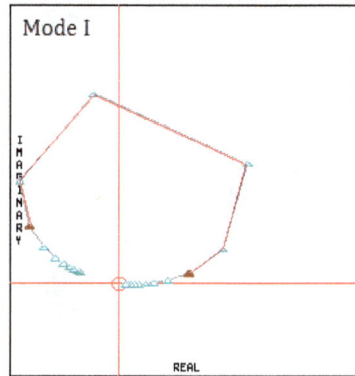

L-FIT FOR MODE I
Natural Frequency (Hz) = 250.37
% Damping (Structural) = 0.8658
Mod. Const. Mag (1/kg) = 10.060
Mod. Const. Phase (o)   = -16.939

(c)

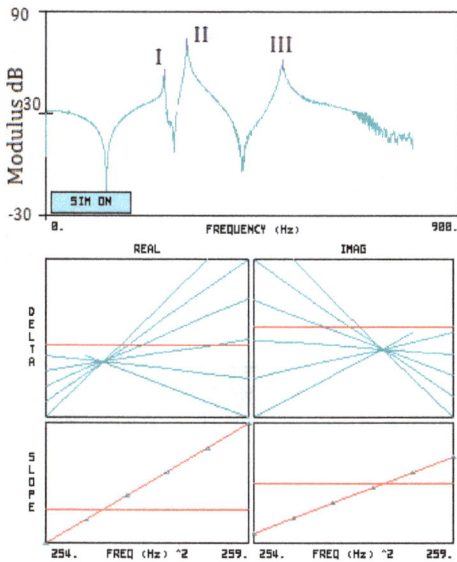

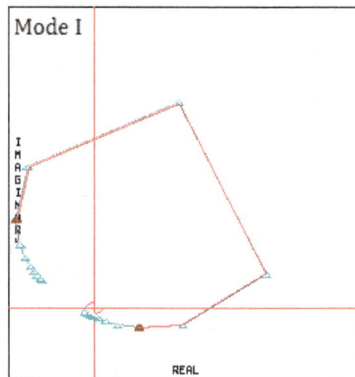

L-FIT FOR MODE I
Natural Frequency (Hz) = 256.25
% Damping (Structural) = 0.7188
Mod. Const. Mag (1/kg) = 3.722
Mod. Const. Phase (o)   = -42.492

(d)

Figure 16.   (*Continued*)

(a)

(b)

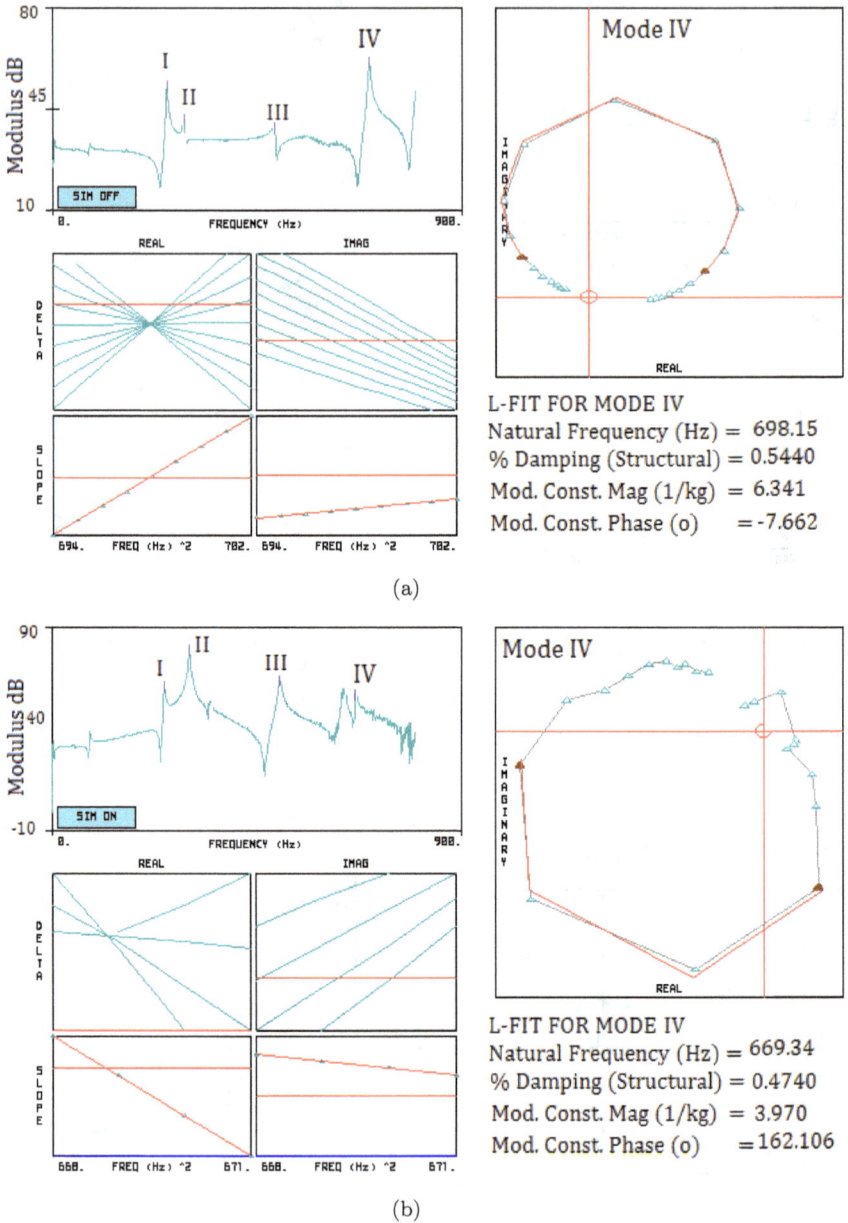

Figure 17.   Cross FRFs and Mode IV LF analyses for (a) pristine, (b) with 5%, (c) with 10%, and (d) with 20% delaminations, respectively.

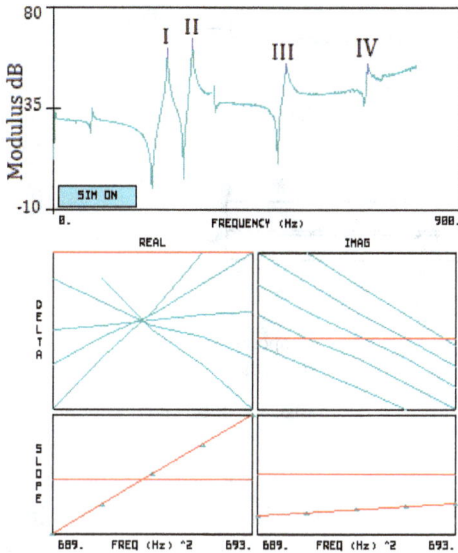

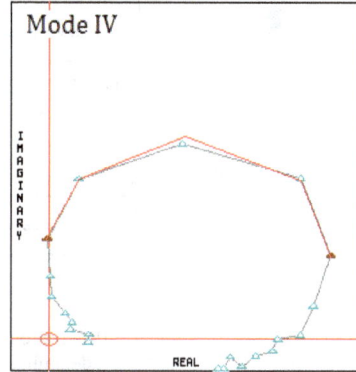

L-FIT FOR MODE IV
Natural Frequency (Hz) = 691.23
% Damping (Structural) = 0.6020
Mod. Const. Mag (1/kg) = 2.613
Mod. Const. Phase (o)   = -11.405

(c)

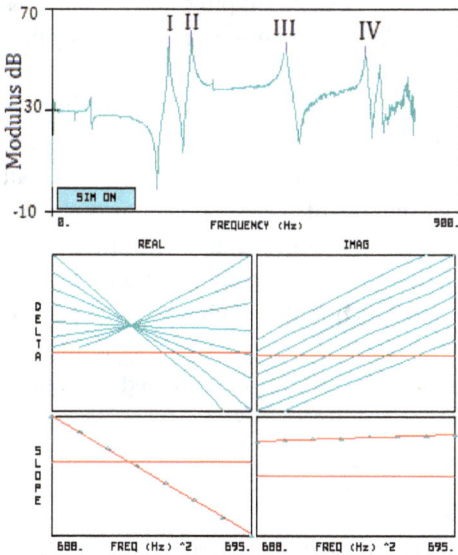

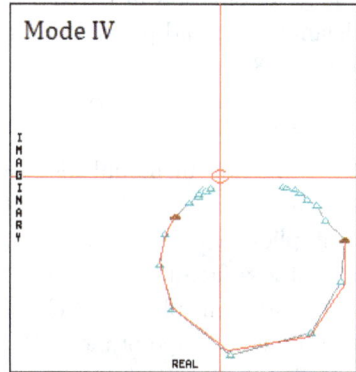

L-FIT FOR MODE IV
Natural Frequency (Hz) = 690.64
% Damping (Structural) = 0.6282
Mod. Const. Mag (1/kg) = 2.902
Mod. Const. Phase (o)   = -177.059

(d)

Figure 17.   (*Continued*)

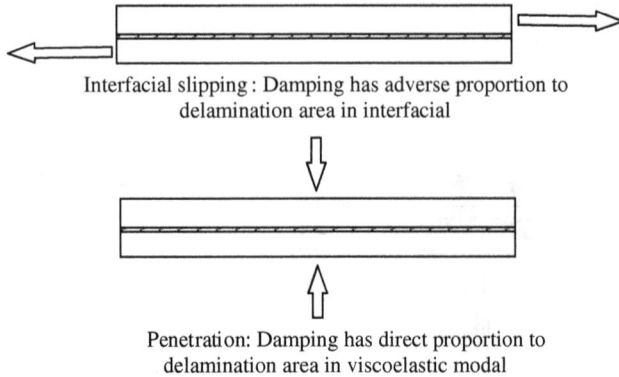

Interfacial slipping : Damping has adverse proportion to
delamination area in interfacial

Penetration: Damping has direct proportion to
delamination area in viscoelastic modal

Figure 18.   Damping mechanisms in composite laminates, i.e. Interfacial slipping and Penetration behaviors.

in the unbound regions between fiber and matrix interface and delamination and (2) damping due to energy dissipation as the result of matrix cracking, fiber breakage etc. Furthermore, Yam *et al.*[24] concluded that energy dissipation in CFRPs is mostly caused by slipping behavior across delamination and penetration tendency between the upper and lower plies, Figure 18.

Based on the observations made in Figures 15–17, and giving the fact that delamination is placed in the mid-plane and at the center of the plate, the reduction in modal loss factors for modes I, II and III which are all bending modes, can be attributed to the loss of penetration between adjacent plies, across the delamination. On the other hand, the increase in modal loss factor associated with mode IV can be associated with sliding of the adjacent plies on each other which is due to the shear effects related to the mode shape of mode IV.

The significant drop observed in modal constants of damaged specimens are also of interest and requires further studies.

6. Conclusion

In this research, a comprehensive experimental investigation has been presented for the analysis of delamination effects on modal parameters of CFRPs. Also, a review of the damping mechanisms in composites is presented and endeavored to justify the study results, using these mechanisms. Modal analysis has been performed based on ELFM. The results prove to be conclusive and able to detect damage in carbon composite laminates as small as 5% of delamination area.

Results show that lower modes' natural frequencies are global characteristics of composite plate and not very suitable for damage detection. On the other hand, modal loss factor has been proved to be a highly sensitive indicator undergoing major changes even at initial damage stages, which makes it appropriate for damage detection and severity estimation. The study reveals conclusive relationship between modal damping change and its associated mode shape. Based on the mode shape, interfacial slipping and/or viscoelastic damping mechanisms are expected to take role in energy dissipation.

The results found imply that tracking damping variations with respect to delamination growth can be used to detect the presence of damage and with some degree of confidence, the controlling damping mechanism(s).

References

1. Amaro, A., Reis, P., and De Moura, M. Delamination effect on bending behavior in carbon-epoxy composites. *Strain* **47**(2), pp. 203–208 (2011).
2. Chandra, R., Singh, S., and Gupta, K. Damping studies in fiber-reinforced composites — A review. *Composite Structure* **46**(1), pp. 41–51 (1999).
3. Zou, Y., Tong, L., and Steven, G. P. Vibration-based model-dependent damage (delamination) identification and health monitoring for composite structures — A Review. *Journal of Sound Vibration* **230**(2), pp. 357–378 (2000).
4. Paolozzi, A. and Peroni, I. Detection of debonding damage in a composite plate through natural frequency variations. *Journal Reinforced Plastics and Composites* **9**(4), pp. 369–389 (1990).
5. Cawley, P. and Adams, R. D. The location of defects in structures from measurements of natural frequencies. *The Journal of Strain Analysis for Engineering Design* **14**(2), pp. 49–57 (2007).
6. Tracy, J. J. and Pardoen, G. C. Effect of delamination on the natural frequencies of composite laminates. *Journal of Composite Materials* **23**(12), pp. 1200–1215 (1989).
7. Kessler, S. S. and Cesnik, C. E. S. Damage detection in composite materials using frequency response methods. *Composites Part B: Engineering* **33**(1), pp. 1–19 (2008).
8. Cao, M. S., Sha, G. G., Gao, Y. F., and Ostachowicz, W. Structural damage identification using damping: A compendium of uses and features. *Smart Materials and Structures* **26**(4), p. 043001 (2017).
9. Kyriazoglou, C., Le Page, B. H., and Guild, F. J. Vibration damping for crack detection in composite laminates. *Composites Part A: Applied Science and Manufacturing* **35**(7–8), pp. 945–953 (2004).

10. Keye, S., Rose, M., and Sachau, D. Localizing delamination damages in aircraft panels from modal damping parameters. In *Proceeding of 19th International Modal Conference (IMAC XIX)*, 2001.

11. Kiral, Z., Murat İçten, B., and Gören Kiral, B. Effect of impact failure on the damping characteristics of beam-like composite structures. *Composites Part B: Engineering* **43**(8), pp. 3053–3060 (2012).

12. Montalvão, D., Ribeiro, A. M. R., and Duarte-Silva, J. A method for the localization of damage in a CFRP plate using damping. *Mechanical Systems Signal Processing* **23**(6), pp. 1846–1854 (2009).

13. Srikanth, N., Kurniawan, L. A., and Gupta, M. Effect of interconnected reinforcement and its content on the damping capacity of aluminum matrix studied by a new circle-fit approach. *Composites Science and Technology* **23**(6), pp. 839–849 (2003).

14. Montalvão, D., Karanatsis, D., Ribeiro, A. M. R., Arina, J., and Baxter, R., An experimental study on the evolution of modal damping with damage in carbon fiber laminates. *Journal of Composite Materials* **49**(19), pp. 2403–2413 (2014).

15. Qiao, P., Lu, K., Lestari, W., and Wang, J. Curvature mode shape-based damage detection in composite laminated plates. *Composite Structures* **80**(3), pp. 409–428 (2007).

16. Pandey, A. K., Biswas, M., and Samman, M. M. Damage detection from changes in curvature mode shapes. *Journal of Sound and Vibration* **145**(2), pp. 321–332 (1991).

17. Tracy, J. J. and Pardoen, G. C. Effect of delamination on the flexural stiffness of composite laminates. *Thin-Walled Structures* **6**(5), pp. 1200–1215 (1988).

18. Goyder, H. G. D. and Harwell, U. Analysis and Identification of Linear and Nonlinear Systems using Random Excitation. *Short Course Notes*, University Manchester (1985).

19. Ewins, D. J. *Modal Testing: Theory, Practice and Application*, 2nd edn. RSP, Philadelphia (2000).

20. He, J. and Fu, Z.-F. *Modal Analysis*, 1st edn. Butterworth-Heinemann, Linacre House, Jordan Hill, Oxford (2001).

21. Kashani, H. and Nobari, A. S. Identification of dynamic characteristics of nonlinear joint based on the optimum equivalent linear frequency response function. *Journal of Sound and Vibration* **329**(9), pp. 1460–1479 (2010).

22. ICATS Manual, Imperial College of Science, Technology and Medicine, Mechanical Engineering Department, Exhibition Road, London, SW7 2BX (1994).

23. Saravanos, D. A. and Hopkins, D. A. Effect of delamination on the damped dynamic characteristics of composite laminates: Analysis and experiments. *Journal of Sound and Vibration* **192**(5), pp. 977–993 (1996).

24. Yam, L. H. Nondestructive detection of internal delamination by vibration-based method for composite plates. *Journal of Composite Materials* **38**(24), pp. 2183–2198 (2004).

25. Treviso, A., Van Genechten, B., Mundo, D., and Tournour, M. Damping in composite materials: Properties and models. *Composites Part B: Engineering* **78**, pp. 144–152 (2015).

26. Bert, C. W. *Composite Materials: A Survey of the Damping Capacity of Fiber-Reinforced Composites*, Oklahoma, 1980.

27. Nelson, D. J. and Hancock, J. W. Interfacial slip and damping in fiber reinforced composites. *Journal of Materials Science* **13**(11), pp. 2429–2440 (1978).

Chapter 4

Problem of Detecting Damage Through Natural Frequency Changes

Gilbert-Rainer Gillich[*,‡], Nuno N. N. Maia[†] and Ion Cornel Mituletu[*]

Department of Mechanical Engineering, University "Eftimie Murgu" of Resita
P-ta Traian Vuia 1–4, 320085 Resita, Romania

†*Department of Mechanical Engineering, University of Lisbon*
IDMEC-Instituto Superior Técnico, Av. Rovisco Pais, 1049–001 Lisbon, Portugal
‡*gr.gillich@uem.ro*

In damage detection, an important aspect is the accurate frequency evaluation which permits observing the modal parameter changes at the earliest stage. The frequency resolution improvement, achieved by extending the time interval of the analysis, is in the case of standard frequency estimation the key to increasing the accuracy of the study. This is not always possible due to the rapid decay of higher-order modes. Many methods to improve frequency readability are nowadays available in the literature, such as those increasing the spectral lines density or those based on several-point interpolation. We debate herein the limitations of the actual methods and introduce an advanced algorithm that uses for interpolation three peaks from three different spectra, achieved from the acquired signal that is stepwise cropped. Processing the signal in this way allows identifying minor natural frequency changes, thus allowing for the observation of damage occurrence in the earliest stage. The algorithm was successfully tested against generated and real signals.

Keywords: Damage detection; Natural frequency; Digital signal processing; Frequency resolution; Interpolation-based methods; Dense overlapped spectrum.

1. Introduction

Damage detection using vibration measurements has gained the attention of numerous researchers and practitioners in the last decades.[1-3] The core of the vibration-based damage detection methods is the existence of a deterministic relation between modal parameters, which can be easily measured, and physical parameters, which have to be estimated.

A common approach in evaluating the structural integrity consists of defining a baseline involving the relevant modal parameters, identified from vibration signals acquired from the intact structure,[4] and all subsequent test results being compared to these data.[5] Any modal parameter change can indicate the occurrence of a structural failure. It becomes possible to assess the location and the magnitude of the failure if the effect of the physical parameter changes is known.[6]

The damage detection methods presented in the literature differ by the level achieved in damage assessment,[7] the complexity of the analyzed structure,[8] the excitation system involved,[9] the type and number of sensors used in the monitoring process,[10] and the techniques used to find out the physical parameter changes from the shifts of the modal parameters. Output-only damage detection methods are among the most promising in engineering applications because of the difficulty in the measurement of excitations. Consequently, it is advantageous to use methods based just on measured responses, where the excitation information is not necessary.[11]

To be considered for real applications, damage detection methods have to demonstrate the capacity to operate well under several limitations. The achievement of accurate results with a limited number of sensors, whose position may not be known in advance, is important. Also important is the capacity to differentiate the modal parameter changes that have occurred due to damage from those resulting from environmental variations or test conditions. A crucial issue is the repeatability of the tests, since a high level of measurement confidence ensures early-stage damage detection. In this context, the natural frequency becomes *a priori* the most suitable modal parameter for damage detection. Nevertheless, methods purely based on frequency shifts require precise frequency evaluation, since frequency changes present low sensitivity to damage.

2. Motivation

Damage detection methods supported by natural frequency changes consist in acquiring, processing, and interpreting vibration signals. The signals acquired in the time domain are converted by specific algorithms into the frequency domain. Most structures have a large enough interval between harmonics, so that discerning between consecutive frequencies is easy. On the other hand, observing incipient damage can be difficult because damage in early states produces limited frequency shifts. The reason is that in standard frequency evaluation, the frequency components of a signal are indicated at lines equidistantly distributed in the spectrum, whose positions are dependent on the signal length. As a consequence, the frequency shift caused by an incipient damage, being small, is not enhanced in the spectrum. This situation persists until the damage grows enough to produce a significant frequency shift and thus the peak-amplitude moves to the next spectral line. Considering this, it is apparent that it is necessary to use advanced algorithms, providing inter-line information regarding the frequency, to increase the accuracy of frequency evaluation.

3. Limitations of the Standard Frequency Evaluation Method

The vibration signals from structures are acquired via transducers as continuous functions of time. In Figure 1, the transducer is exemplified by a piezoelectric accelerometer that produces an analog electrical output proportional to the structure acceleration at the measurement point. The electrical output is sent from the transducer to an analog-to-digital (A/D) converter that records the signal amplitude at discrete time intervals. This operation is known as sampling.

The continuous signal, see Figure 2, is transformed in this way into a digital signal that is expressed as a discrete time series.

Figure 3 illustrates the resulted sampled signal processed by the computer. The time interval between two consecutive instants, called sampling time or time resolution, is derived from the sampling rate F_S as

$$\tau = \frac{1}{F_S} \tag{1}$$

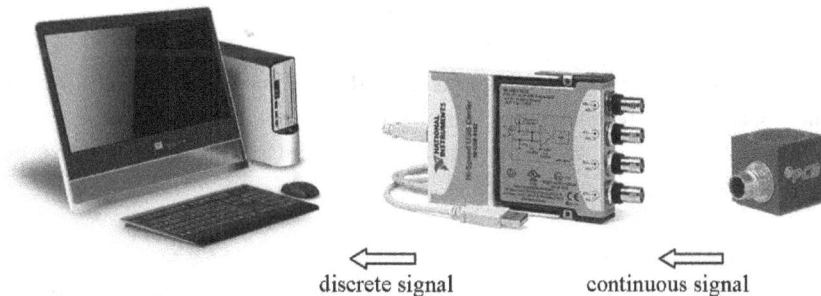

discrete signal   continuous signal

Figure 1.   The conversion of a continuous signal to a digital code.

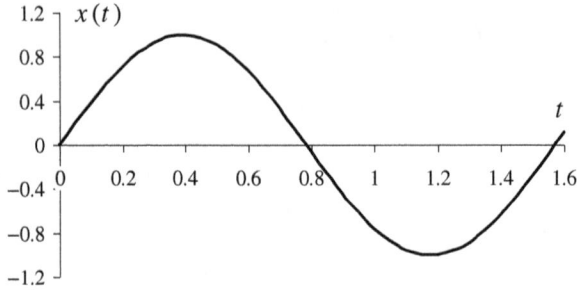

Figure 2. The continuous measured signal.

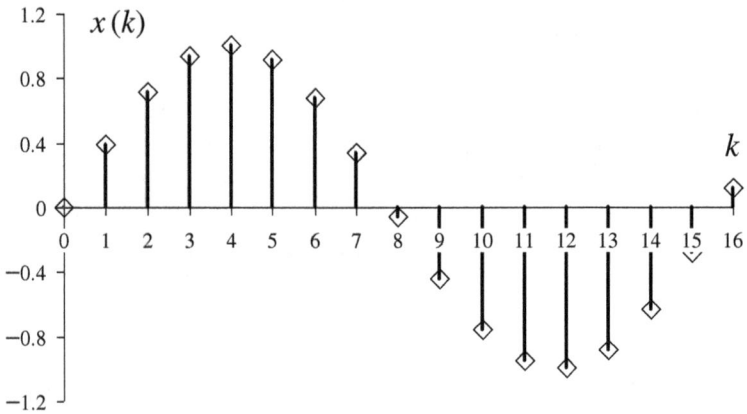

Figure 3. The equivalent digital signal resulted by conversion.

For the *NI 9233* module shown in Figure 1, the sampling rate, in Hz, is[12]

$$F_S = \frac{50000}{m} \qquad (2)$$

with m as an integer less than 25.

Digitization of the continuous signal $x(t)$ means ensuring a value $x[k] = x(k\tau)$ representing the amplitude to each discrete time $k\tau$, with $k = 0, 1, 2, \ldots, N-1$. As a consequence, the signal $x(t)$ achieves the discrete representation as a sequence

$$\{x\} = \{x[0], x[1], \ldots, x[k], \ldots, x[N-1]\} \qquad (3)$$

The sequence $\{x\}$ contains no explicit information about the sampling rate FS. Nevertheless, because the sampling time τ and the number of samples N are known, the signal time length is derived as

$$T_S = (N-1)\tau = \frac{(N-1)}{F_S} \tag{4}$$

The sampling rate F_S and subsequently the sampling time τ define the quality of an A/D conversion. In accordance with the Nyquist–Shannon sampling theorem, a proper signal reconstruction is possible if the highest signal frequency f_H satisfies the condition

$$f_H < \frac{F_S}{2} \tag{5}$$

Converting a sinusoid of amplitude a and frequency f into a discrete signal gives

$$x(t) = a\sin(2\pi f t) \rightarrow x[\tau] = x(k\tau) = a\sin(2\pi f k\tau) \tag{6}$$

Any signal can be composed employing a set of harmonics. It is usual to nominate such a signal as a multitone signal. On the other hand, a multitone signal can be decomposed in order to highlight its harmonic components. A widely used algorithm to represent a function in the frequency domain is the Discrete Fourier Transform (DFT). Hence, the sequence $\{x\}$ indicated in Equation (3) is expressed as a sum of sinusoids

$$x[k] = \sum_{j=0}^{N-1} a_j e^{i2\pi \frac{k}{N-1} j} \tag{7}$$

where j indicates the sinusoid index and $i^2 = -1$.

By definition, the coefficients a_j are defined as

$$a_j = \frac{1}{N} \sum_{j=0}^{N-1} x[k] e^{-i2\pi \frac{k}{N-1} j} \tag{8}$$

Reconstructing the signal from the sequence $\{x\}$ given in Equation (3), a continuous time function $x(k\tau) = \hat{x}(t)$ is obtained, which is an approximation of the original continuous time function $x(t)$. The higher the ratio F_S/f_H, the better the approximation.

The continuous version of the discrete function $x[k]$ is

$$x(k\tau) = \sum_{j=0}^{N-1} a_j e^{i2\pi \frac{k}{N-1} j} \tag{9}$$

for $x \in R$, or involving the time as independent parameter

$$\widehat{x}(t) = \sum_{j=0}^{N-1} a_j e^{i2\pi \frac{j}{N-1} \frac{t}{\tau}} = \sum_{j=0}^{N-1} a_j e^{i2\pi f_j t} \tag{10}$$

where f_j denotes the frequency of the jth component, defined by

$$f_j = \frac{j}{(N-1)\tau} = \Delta f \cdot j \tag{11}$$

Equation (11) automatically results in

$$\Delta f = \frac{1}{(N-1)\tau} = \frac{1}{T_S} \tag{12}$$

Equation (11) shows that the frequencies f_j, where the amplitudes are indicated, have equidistantly distributed positions in the spectrum and are known as spectral lines. The distance Δf between two consecutive spectral lines is the frequency resolution. Regarding the amplitudes, the DFT of the sequence $\{x\}$ indicates, at the jth frequency, the amplitude

$$X[f_j] = \sum_{k=0}^{N-1} x[k] e^{-i2\pi f_j t} \tag{13}$$

If spectral line numbers j are considered in standing up of frequencies f_j, Equation (13) becomes

$$X[j] = N a_j = \sum_{k=0}^{N-1} x[k] e^{-i2\pi \frac{j}{N-1} k} \tag{14}$$

A signal having N samples in the time domain results in a spectrum with N spectral lines, as illustrated in Figures 4 and 5. The amplitude values $X[j] = X[f_j]$ are indicated at the $j = 0, 1, 2, \ldots, N - 1$ spectral lines.

For signals as those acquired from the output of an A/D converter, the DFT-based spectrum is symmetric and has the following property

$$|X[j]| = |X[N - j]| \tag{15}$$

Therefore, just half of the spectral lines contain useful information, so only those need to be computed; the other spectral lines contain redundant information because of the symmetry. Accordingly, only $N/2$ spectral lines

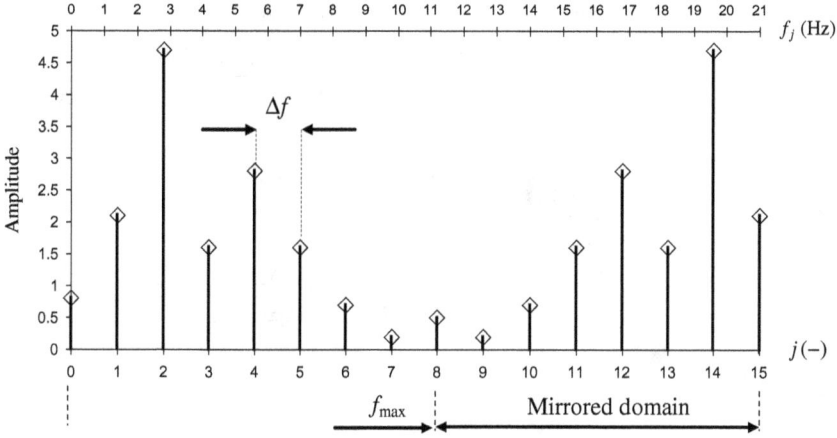

Figure 4. DFT-based spectrum of a sampled signal — even number of samples N.

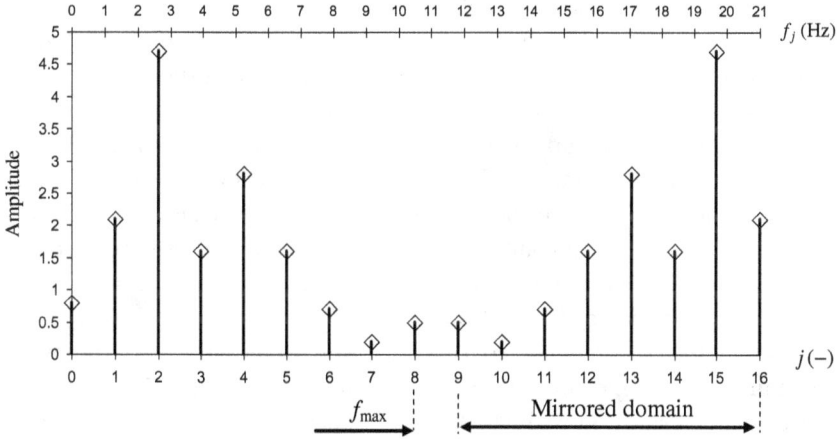

Figure 5. DFT-based spectrum of a sampled signal — odd number of samples N.

in the DFT need to be analyzed, the highest frequency being the so-called Nyquist frequency that is

$$f_{Ny} = \frac{N}{2}\Delta f \approx \frac{1}{2\tau} = \frac{F_S}{2} \tag{16}$$

However, the Nyquist frequency f_{Ny} does not provide a precise estimate regarding the highest frequency f_{max} that can be derived. It depends on the nature of N, which can be an even or odd number. A precise estimation of f_{max} is made based on the property stated in Equation (15), the results

Table 1.   The maximum achievable frequency and the spectral line distribution.

	N: Even number of samples		N: Odd number of samples	
Index number	Index value j	Frequency f_j	Index value j	Frequency f_j
1	0	DC	0	DC
2	1	Δf	1	Δf
3	2	$2\Delta f$	2	$2\Delta f$
...	...	...	...	...
$j+1$	$\frac{N}{2}$	$\pm\frac{N}{2}\Delta f = f_{\max}$	$\frac{N-1}{2}$	$\frac{N-1}{2}\Delta f = f_{\max}$
...	...	...	$\frac{N-1}{2}+1$	$-\frac{N-1}{2}\Delta f$
...	...	...	...	...
$N-1$	$N-2$	$-2\Delta f$	$N-2$	$-2\Delta f$
N	$N-1$	$-\Delta f$	$N-1$	$-\Delta f$

being presented in Table 1. Here also, the spectral line distribution and the associated frequencies are presented.

Because in the DFT-representation the frequencies are indicated on predefined spectral lines, whose positions are set through the analyzed time length of the signal, these are not necessarily the true frequencies contained in the signal. An error up to the half frequency resolution is possible, this being the main disadvantage of the standard frequency estimators. A long analysis time assures dense spectral lines and in consequence, an acceptable frequency readability. In contrast, short-time signals as those that occur in damage detection lead to a weak frequency resolution and possible important errors in frequency estimation.

Hereinafter, we demonstrate that the results achieved, whenever involving standard frequency evaluation, depend on the acquisition strategy. The test signal is a sinusoid with the known frequency $f = 5\,\text{Hz}$ and amplitude $A = 1\,\text{mm/s}^2$. It is generated with the sampling frequency $F_S = 100\,\text{Hz}$, resulting in a sampling time $\tau = 0.01\,\text{s}$. First, the DFT is applied to a segment consisting of a sequence of $N_1 = 101$ samples, to which corresponds the time length $T_{S1} = 1\,\text{s}$. This segment, nominated as signal 1, contains an integer number of periods and has the last sample marked with a gray square in Figure 6. In consequence, the DFT plotted with a black line in Figure 7 indicates the correct frequency. The amplitude is equal to that of the sinusoid, thus the value $1\,\text{mm/s}^2$ is expected. As Figure 7 illustrates the real DFT, the spectral representation is single-sided and only the positive spectral lines are visible. At the spectral line with $f = 5\,\text{Hz}$ the achieved amplitude is indeed $1\,\text{mm/s}^2$.

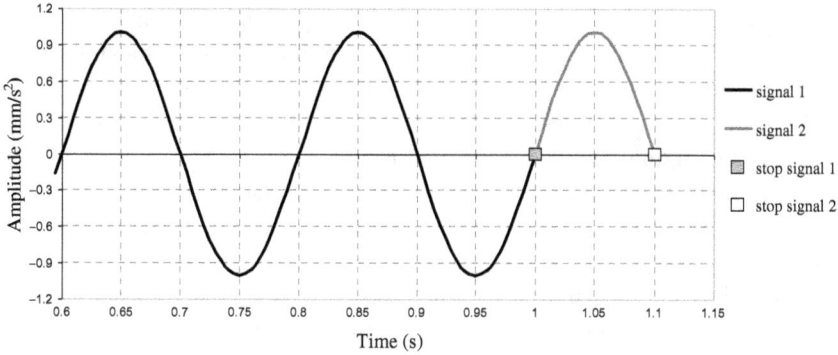

Figure 6.   Harmonic signal with integer number of periods (last sample gray square) and non-integer number of periods (last sample white square), respectively.

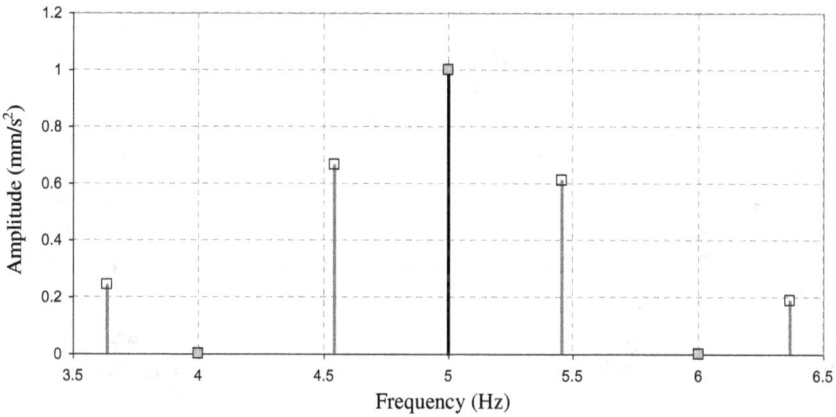

Figure 7.   DFT of a sinusoidal signal with integer number of periods (gray top square) and non-integer number of periods (white top square), respectively.

The second analysis concerns a signal segment with time length $T_{S2} = 1.1\,\text{s}$, whose discrete sequence contains $N_2 = 111$ samples, the last of them being indicated with a white square in Figure 6. This segment contains a non-integer number of periods and therefore the DFT, plotted in Figure 7 with gray lines marked at the top with white squares, indicates false frequencies ($4.545\,\text{Hz}$ instead of $5\,\text{Hz}$). Also the peak amplitude, indicated as $0.6659\,\text{mm/s}^2$, does not reflect the expected one. So, neither the amplitude nor the frequency is correctly indicated. The error occurs in spite of the increased frequency resolution. In fact, the position of the spectral line is important and it should match the frequency contained in the signal.

Otherwise, the DFT algorithm considers that the signal contains harmonics having as periods multiples of the frequency resolution, and indicates amplitudes at certain lines in the spectrum. Obviously, if the analyzed signal is a sinusoid, the amplitudes spread along the spectral lines are smaller than that of the original signal.[13,14] In this case, the peak amplitude is accomplished on the spectral line that is nearest to the real frequency, where the estimated frequency is found. The biggest difference that can occur between the estimated frequency and the real frequency is half of the frequency resolution. Since the maximum possible error $\varepsilon_{max} = \Delta f/2$ can be predicted, the actual error cannot be predicted because it depends on the signal period $T = 1/f$ which is obviously not known *a priori*.

We can improve the frequency readability by employing several simple techniques. The simplest attempt consists in extending the observation time T_S, which makes the spectral lines denser and causes a finer frequency resolution. Figure 8 presents the harmonic signal that has the frequency $f = 5\,\text{Hz}$ and the amplitude $A = 1\,\text{mm/s}^2$. The signal was generated with the sampling frequency $F_S = 100\,\text{Hz}$.

The original signal has the length $T_{S2} = 1.1\,\text{s}$, whereas the extended signals have the randomly chosen lengths $T_{S3} = 1.62\,\text{s}$ and $T_{S4} = 2.1\,\text{s}$. The signal settings are given in Table 2 and the results accomplished by applying the DFT are indicated in Table 3 and depicted in Figure 9.

The DFT of the original signal indicates the frequency $f_2 = 4.54545\,\text{Hz}$, which differs from the frequency generated with $f = 5\,\text{Hz}$, the resulting error being $\varepsilon_2 = 0.45455\,\text{Hz}$. On the other hand, the frequencies extracted from the signals with extended observation times are $f_3 = 4.98327\,\text{Hz}$ and

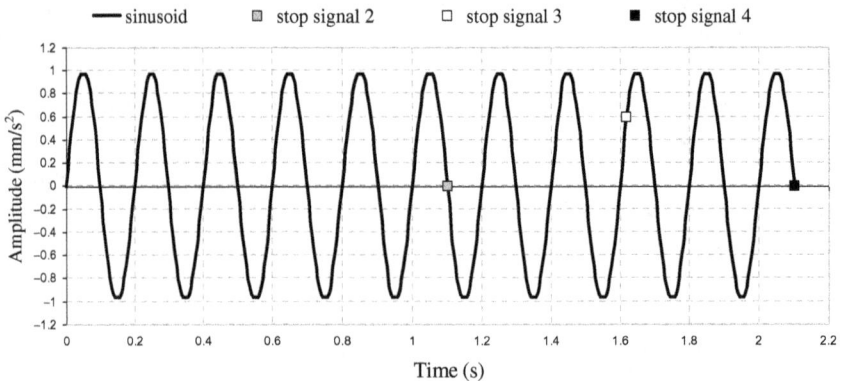

Figure 8. The original signal and its extensions due to increasing observation time.

Table 2.   Settings for generating the analyzed signals.

Signal name	Parameters			
	T_S (s)	N	F_S	Δf (Hz)
Signal 2	1.1	111	100	0.909091
Signal 3	1.62	163	100	0.617284
Signal 4	2.1	211	100	0.47619

Table 3.   Frequencies and amplitudes achieved by standard evaluation.

Signal name	f (Hz)	A_{DFT} (mm/s^2)	ε (Hz)	ε (%)
Signal 2	4.54545	0.66597	0.45455	9.091
Signal 3	4.93827	0.97772	0.06173	1.234
Signal 4	4.76190	0.65164	0.23810	4.762

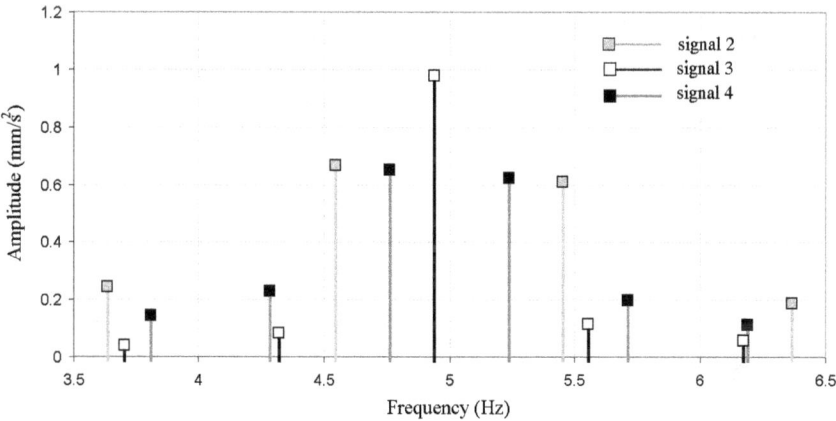

Figure 9.   DFT of signals with different observation times.

$f_4 = 4.7619\,\text{Hz}$, respectively, as indicated in Table 3. Both of them are more precise than those found for the original signal 2, the errors being reduced to $\varepsilon_3 = 0.06173\,\text{Hz}$ and $\varepsilon_4 = 0.23810\,\text{Hz}$, respectively. Maybe unexpectedly, the longest signal does not ensure the best result. This test shows that even if increasing the observation time ensures a finer frequency resolution and in consequence reduces the maximum possible error, it does not guarantee an improvement in the frequency estimation.

It is impossible to increase the observation time for signals acquired from the free vibration of damped systems. To overcome this, the procedure known as zero-padding can be applied. It consists of lengthening the signal by adding numerous samples N_{ZP} with null amplitude.[15] It does not change the original sampling rate and the frequency range of the DFT output remains the same. Because the increased numbers of the resulting spectral lines are now evenly distributed over the same frequency domain, from 0 to $F_S/2$, the spacing between lines must decrease to fit more samples over the same frequency range. Due to zero-padding, the frequency resolution increases from

$$\Delta f = \frac{F_S}{N-1} \tag{17}$$

as it is for the original signal, to

$$\Delta f_{ZP} = \frac{F_S}{N + N_{ZP} - 1} \tag{18}$$

The zero-padded signal is shown in Figure 10. This signal has the length $T_{S-ZP} = 2.2\,\text{s}$ and consists of a sinusoid with the length $T_{S2} = 1.1\,\text{s}$ that has appended a DC component with the same length as the sinusoid and with zero amplitude. The settings used to generate the zero-padded signal are given in Table 4 and the achieved results are presented in Table 5.

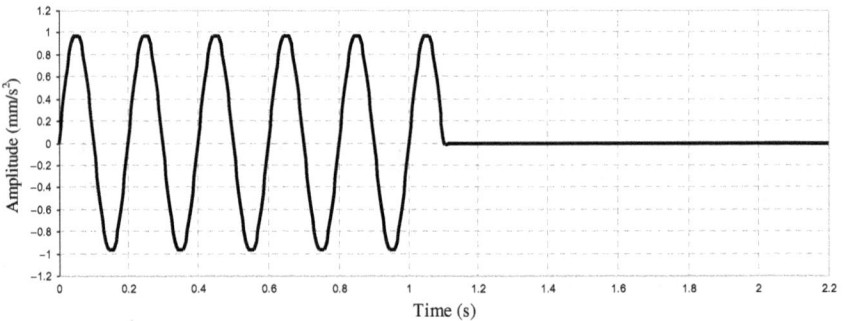

Figure 10. The original signal lengthened by zero-padding.

Table 4. Settings for generating the zero-padded signal.

Signal name	T_S (s)	N	F_S	Δf (Hz)	A (mm/s²)
			Parameters		
Signal 2	1.1	111	100	0.909091	1
Signal DC	1.1	111	100	0.909091	0
Signal ZP1	2.2	222	100	0.454545	—

Table 5.   Frequencies and amplitudes achieved by standard evaluation.

Signal name	f (Hz)	A_{DFT} (mm/s^2)	ε (Hz)	ε (%)
Signal 2	4.54545	0.66597	0.45455	9.091
Signal ZP1	5	0.5	0	0

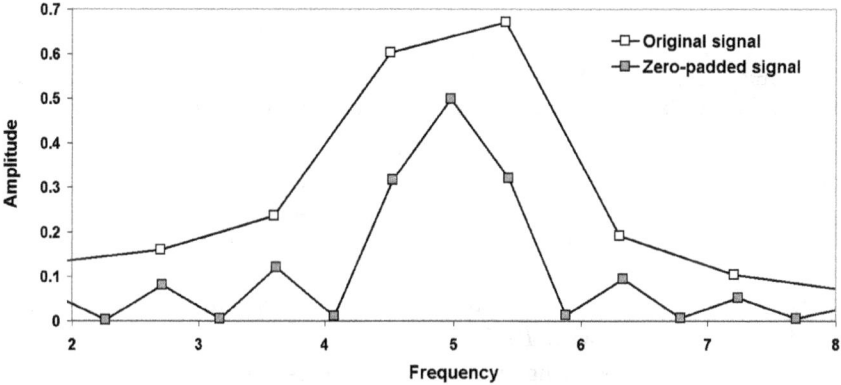

Figure 11.   DFT representation of the original and the zero-padded signals. Note the lower magnitude and higher precision achieved by zero-padding.

A comparison between DFTs obtained for the signal and the zero-padded signal, respectively, is made in Figure 11. Observing this figure we note that for an artificially increased number of samples the peak amplitude in the spectrum decreases. In some cases, zero-padding becomes inefficient. On the other hand, the number of spectral lines increases and therefore the distance between two consecutive spectral lines decreases. Due to this advantage, the biggest possible error $\varepsilon_{\max}$ is diminished and the probability to improve the frequency evaluation results is obvious. In this case, the real frequency was found by chance.

Because zero-padding is an interpolation, the denser spectral lines do not have a real frequency resolution increase. Zero-padding reduces neither leakage nor the main lobe width so two closely located frequency components of a signal are not observable if zero-padding is applied. For exemplification, the DFT of signal 4, described in Table 2, is employed. It is represented by gray squares in Figure 12, where it can be recognized as having the biggest distance between two spectral lines. By extending the time length with T_S-$ZP2 = 1.82$ s by zero-padding, the spectral lines

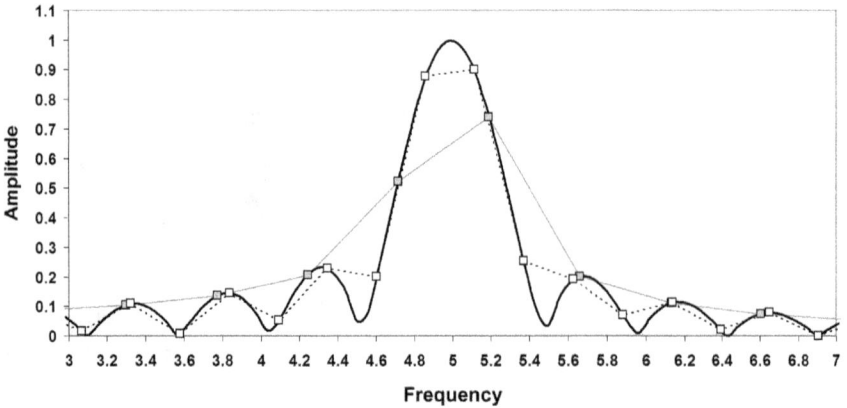

Figure 12.   DFT representation of signal 4 (gray squares) and of two signals resulted after zero-padding with the additional time 1.8 s (white squares) and 15 s (black points). For the zero-padded signal, the scaling was with factors 1.87 and 8.14. respectively.

become closer. The DFT is represented by white squares in Figure 12. If the time is extended with T_S-$ZP3$ = 15 s, the spectral lines become close enough to permit observing that the envelope is a *sinc* function. For this time length, a precise evaluation of the generated frequency of the signal is possible. Because the zero-amplitude samples introduce no energy, the amplitude of those signals in the frequency domain decreases dramatically. To make the DFTs in Figure 12 comparable, we scaled the amplitudes of the zero-padded signals by a factor derived with the relation

$$\eta = \frac{N}{N + N_{ZP}} \tag{19}$$

where N is the number of samples of the analyzed signal and N_{ZP} is the number of samples of the signal extension. Figure 12 shows that, however long the time added to the original signal is, the lobes have the same width. An increased number of samples lead to an increased number of points in a lobe, without increasing the real frequency resolution. This is observable in Figure 13, where the DFTs, derived from two signals, are depicted. The first signal results from the superposition of two sine waves having the frequencies $f_1 = 5$ Hz and $f_2 = 5.2$ Hz, both achieving the amplitude $A = 1$ mm/s². The number of employed samples is $N = 2100$ and the sampling rate $F_S = 100$ Hz. The second signal is a shorter version of the first signal, having just $N_R = 210$ samples. This signal is zero-padded with $N_{ZP} = 1890$ samples in order to achieve the same time length as the first signal. In this way, the same spectral line distribution is

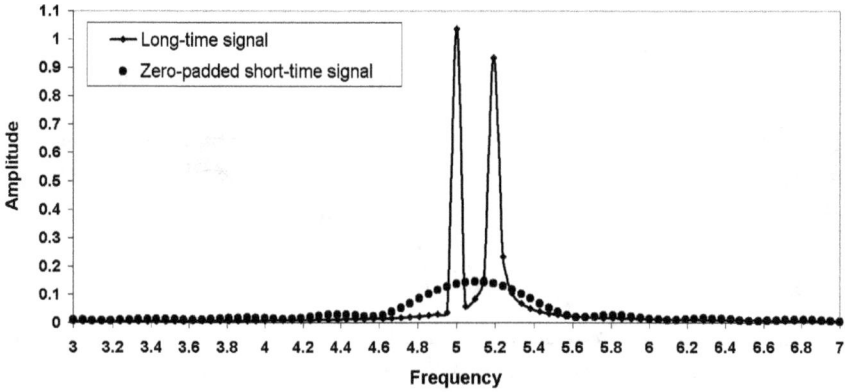

Figure 13. DFT representation of a signal containing 2 sinusoids, achieved by standard frequency evaluation of a long-time signal (continuous line) and similar short-time signal after zero-padding (dotted line).

attained even though a significant difference between the two spectra can be observed.

From Figure 13, it is clear that the DFT of a signal acquired (or generated) in a large time interval, marked here with a continuous line, has a fine frequency resolution and permits identification of closely located frequency components. In this example, the frequencies $f_1 = 5\,\text{Hz}$ and $f_2 = 5.2\,\text{Hz}$ have separate lobes and are easily recognizable. This is not the case of the zero-padded signal, plotted with a dotted line in Figure 13. Despite an apparently good frequency resolution, one lobe covers both spectral lines and makes it impossible to identify the two components. As a consequence, the estimated frequency is wrongly found between the two real frequency components. Moreover, the pick amplitude is dramatically diminished.

Signal windowing is also considered as an alternative to improve the frequency readability. When performing spectral analysis on finite-length time signals containing a non-integer number of cycles, the multiplication with a windowing function reduces the amplitudes of the signal ends, making them decrease gradually towards zero.[16,17] The linearly represented DFT gets no improvement, as shown in Figure 14, although the discontinuities of the signal edges are minimized, reducing spectral leakage in the DFT expressed in dB (see Figure 15). As neither the frequency resolution is improved nor the spectral lines are repositioned, the frequency estimation cannot be improved.

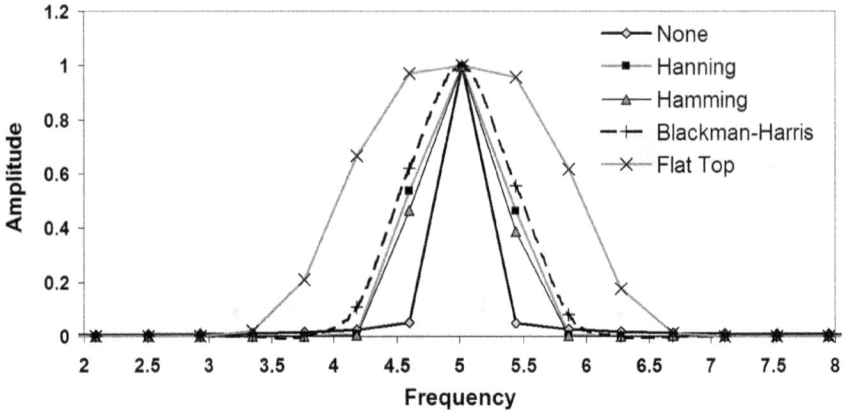

Figure 14. Linear DFT representation of a 5-Hz sinusoid in the absence and presence of windowing.

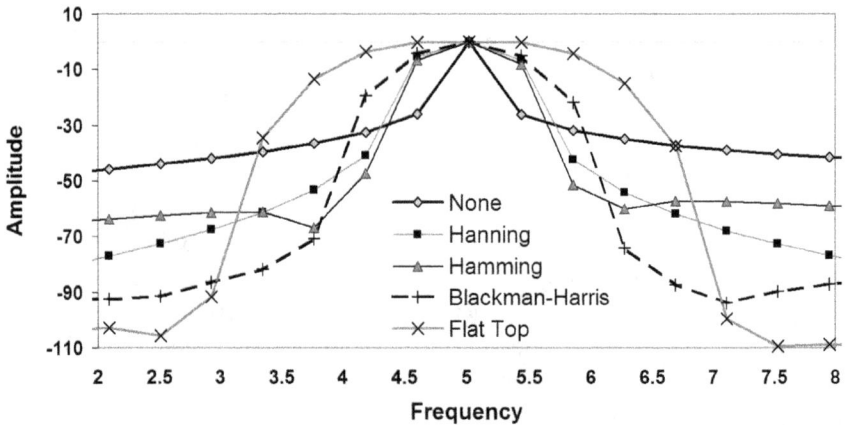

Figure 15. Logarithmic DFT representation of a 5-Hz sinusoid in the absence and presence of windowing.

In accordance with the presented experiments, we can conclude that extending the observation time is the only method that permits accurate frequency evaluation for signals with one or multiple frequency components. As this is not always possible, an alternative is to artificially increase the observation time by adding a sequence of zeros to the original signal. By this method, the biggest possible error is reduced by decreasing the distance between the spectral lines but the main lobe width is not reduced and hence is not able to find closely located frequencies.

The DFT in the dB representation of a windowed signal can indicate the existence of closely located frequencies without improving the frequency estimation. None of the previously presented methods ensures sufficient precision for frequency evaluation of short-time signals that commonly result in the damage detection process. This makes the investigation of other methods necessary.

4. Interpolation-Based Methods to Improve the Frequency Readability

In damage detection, to improve the frequency readability, the interpolation methods are preferred because these are simple and require low computational resources. Moreover, the natural frequencies are seldom closely located, excepting the case of plates. Actual interpolation methods are based on analyzing two or three points belonging to a spectrum derived from vibration signal measurements. In this section several methods are analyzed and tested against an advanced method that considers three points belonging to three separate spectra obtained from one signal which is stepwise cropped.

4.1. *Interpolation methods based on analyzing one spectrum*

The frequencies evaluated with standard methods provide results that are directly linked to the position of the spectral lines, as we can observe in Figure 16. Finding the real frequency at an inter-line position relies on plotting the curve best fitting to several points obtained in the DFT. Let us consider one of the three points being the maximizer, which has as coordinates the peak amplitude A_j and its associated spectral line index j. Its two neighboring spectral lines $j - 1$ and $j + 1$ indicate the amplitudes A_{j-1} and A_{j+1}. The real frequency is near the maximizer, at the inter-line position j_{real}; here the amplitude A of the harmonic signal is expected. The idea is to find a fractional correction term δ that is the distance between the maximizer position j and the inter-line position j_{corr} where the interpolation curve reaches the highest amplitude A_{max}. Obviously, j_{real} and j_{corr} should be as close as possible.

Once the correction term is known, the corrected frequency f_{corr} is found by adjusting the read frequency $f_j = j\Delta f$ with the fraction of the frequency resolution $\Lambda f = \delta\Delta f$. The estimation of displacement δ is referred to as the *fine frequency estimation*, as opposed to the *coarse frequency estimation* performed by locating DFT maximum.[18]

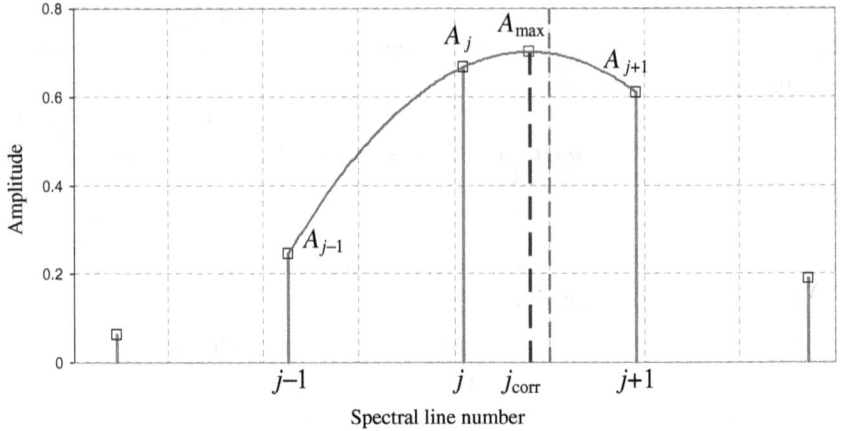

Figure 16.   DFT output for a sinusoid (frequency $f = 5$ Hz) by standard evaluation.

Grandke presented an efficient method[19] that involves the DFT peak A_j and its largest neighbor, the DFT being achieved from the time domain signal to which a Hanning window is previously applied. If in the spectral representation the maximizer largest neighbor is A_{j-1}, the correction term δ results from calculating the following ratios:

$$\alpha = \frac{A_{j-1}}{A_j} \tag{20}$$

$$\delta = \frac{2\alpha - 1}{\alpha + 1} \tag{21}$$

The corrected frequency results from

$$f_{\text{corr}} = (j + \delta)\Delta f \tag{22}$$

If the maximizer largest neighbor is A_{j+1} as in Figure 16, the correction term results from

$$\alpha = \frac{A_{j+1}}{A_j} \tag{23}$$

and the corrected frequency is

$$f_{\text{corr}} = (j + 1 + \delta)\Delta f \tag{24}$$

A similar method was developed by Quinn, but it utilizes the DFT output maximizer and both neighbor amplitudes.[20] However, two interpolations are employed, each involving just two amplitudes. The method is fast because it requires no windowing of the signal before the DFT is applied. Following ratios are calculated for Quinn's estimator:

$$\alpha_1 = \frac{A_{j-1}}{A_j} \tag{25}$$

$$\delta_1 = \frac{\alpha_1}{1 - \alpha_1} \tag{26}$$

and

$$\alpha_2 = \frac{A_{j+1}}{A_j} \tag{27}$$

$$\delta_2 = \frac{-\alpha_2}{1 - \alpha_2} \tag{28}$$

The corrected frequency f_{corr} is calculated with Equation (22), the correction term being chosen as follows:

- if $|\delta_1| > |\delta_2|$ then $\delta = \delta_2$
- else $\delta = \delta_1$.

Another similar interpolation method[21] is proposed by Jain *et al.* Again, the amplitudes on the neighbor lines of the maximizer define the terms in which the frequency correction estimate is calculated. If $A_{j-1} > A_{j+1}$ then

$$\alpha_1 = \frac{A_j}{A_{j-1}} \tag{29}$$

$$\delta_1 = \frac{\alpha_1}{1 + \alpha_1} \tag{30}$$

and the corrected frequency is derived with the relation

$$f_{\text{corr}} = (j - 1 + \delta_1)\Delta f \tag{31}$$

Else, the correction term δ results from

$$\alpha_2 = \frac{A_{j+1}}{A_j} \tag{32}$$

$$\delta_2 = \frac{-\alpha_2}{1 - \alpha_2} \tag{33}$$

and the corrected frequency is derived involving Equation (22).

The next interpolation algorithms make use of three amplitudes at once. Ding proposed a barycentric method,[22] where the correction term δ

results from

$$\delta = \frac{A_{j+1} - A_{j-1}}{A_{j-1} + A_j + A_{j+1}} \tag{34}$$

Another correction term, which involves a quadratic method, is presented by Voglewede.[23] It is found from the relation

$$\delta = \frac{A_{j+1} - A_{j-1}}{2(2A_j - A_{j-1} - A_{j+1})} \tag{35}$$

Instead of this quadratic estimator, Jacobsen proposes a new quadratic estimator,[24] which reads

$$\delta_{Jac} = \frac{A_{j+1} - A_{j-1}}{2A_j - A_{j-1} - A_{j+1}} \tag{36}$$

Çandan,[25] proposed a relation to improve Jaconsen's correction estimator, that is,

$$\delta = \frac{\tan(\pi/N)}{(\pi/N)} \delta_{Jac} \tag{37}$$

For the correction estimates derived involving Equations (32)–(37), the corrected frequency is calculated using Equation (22).

4.2. Interpolation methods based on analyzing three spectra

The weak point of the above-presented frequency estimators based on interpolation is the need for windowing, which requires time and computational resource availability. Windowing the signal before applying the DFT is necessary, otherwise, just two points in the spectrum belong to the main lobe, see Figures 14 and 15, and using three points for interpolation makes no sense. Processing the signal by zero-padding before interpolation is also an available option, even if not mentioned in the literature. Doubling the number of samples by zero-padding ensures four points in the main lobe, this way an efficient interpolation is possible. The layout of the DFT outcomes for the original and the zero-padded signal is illustrated in Figure 17.

An advanced frequency estimation method, which does not imply windowing or zero-padding, is presented in what follows. It is based on the observation that different peak amplitudes are indicated around the real frequency f_R when different analysis times are employed. Therefore, we can use three peak amplitudes belonging to three different spectra of the same signal to perform the interpolation and find the real frequency.

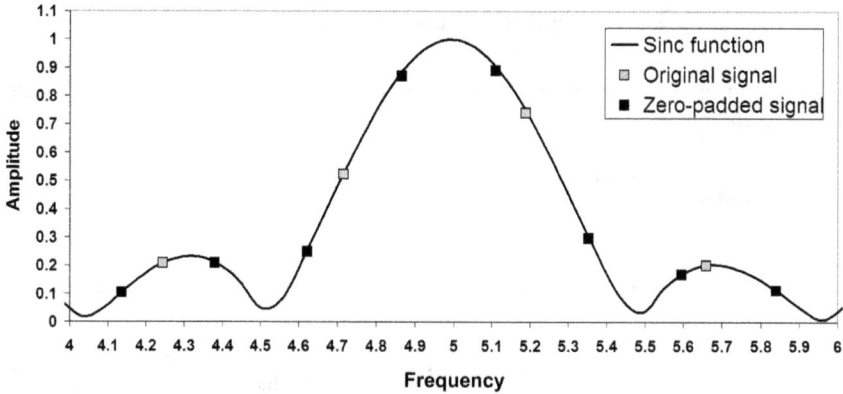

Figure 17. DFT representation of the original signal (white squares) and of the resulted signal after 1:2 zero-padding (black squares). The main lobe and the side-lobes are plotted using the absolute *sinc* function.

The three spectra are obtained from the iteratively cropped time domain signal. The algorithm consists of applying the steps presented below:

(1) Perform a coarse frequency estimation, which can be made by a standard evaluation employing the DFT. It results in the estimated frequency $f_{E\text{-prim}}$.

(2) Calculate the period corresponding to the coarse estimated frequency with the relation.

$$T_{E\text{-prim}} = 1/f_{E\text{-prim}} \tag{38}$$

(3) Establish the initial analysis time length

$$T_{S\text{-prim}} = (n + 0.45)T_{E\text{-prim}} \tag{39}$$

where n is the number of integer periods $T_{E\text{-prim}}$ contained in the initial analysis time.

(4) Calculate the length of the shortest signal employed in the analysis.

$$T_{S\text{-fin}} = (n - 0.45)T_{E\text{-prim}} \tag{40}$$

(5) Calculate the time resolution

$$\tau = \frac{T_S}{N_S - 1} \tag{41}$$

where T_S is the acquisition time and N_S is the number of samples contained in the acquired signal.

(6) Calculate the number of samples contained in the initial and final analyzed signals.

$$N_{S\text{-prim}} = \frac{T_{S\text{-prim}}}{\tau} + 1 \quad \text{and} \quad N_{S\text{-fin}} = \frac{T_{S\text{-fin}}}{\tau} + 1, \quad \text{respectively} \quad (42)$$

(7) Calculate the total number of samples to be reduced by cropping the signal.

$$N_{S\text{-red}} = N_{S\text{-prim}} - N_{S\text{-fin}} \tag{43}$$

(8) Define the number of proposed iterations μ.
(9) Calculate the numbers to be reduced by one iteration.

$$N_{S\text{-it}} = \frac{N_{S\text{-prim}} - N_{S\text{-fin}}}{\mu - 1} \tag{44}$$

(10) Round down $N_{S\text{-it}}$ to obtain an integer number.
(11) Define the initial signal by cropping the original acquired signal in order to have contained $N_{S\text{-prim}}$ samples.
(12) Iteratively crop the initial signal with $N_{S\text{-it}}$ samples by iteration.
(13) Apply the DFT to the $\mu + 1$ achieved signals.
(14) Overlap the achieved spectra.
(15) Select the maximizer and its neighbor amplitudes.

To apply this algorithm appears difficult, but in fact it is not the case since the analysis implies setting of just two parameters, the number of iterations and the frequency range framing the analyzed harmonic component, all other operations being performed by the computer.

Example of an overlapped spectrum is given in Figure 18. Interpolation is not necessary if the number of iterations is significant because the spectral line on which the peak amplitude $A_{\max}$ in the overlapped spectrum is found, i.e. f_{corr}, is close to the real frequency.[26] A disadvantage should be the repetitive application of the DFT, which requires the availability of computational resources, but the performance of actual computers outgrows the problem.

Note that the total number of samples in the acquired signal N_S has no influence on the frequency resolution, but ensures a large number of samples

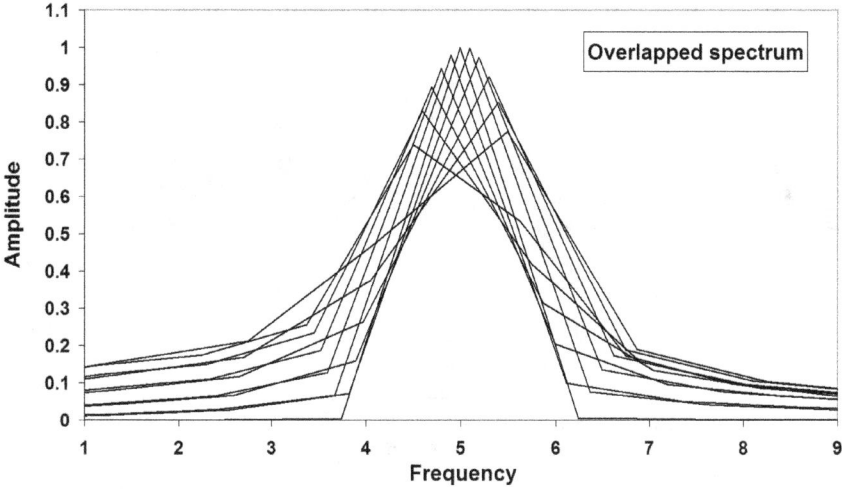

Figure 18.   Overlapped spectrum achieved by plotting several DFTs in one spectral representation.

to the time interval constituting one period T_E. Because the number of samples $N_{S\text{-red}}$ to be cut from the initial sequence is related to the period T_E, the higher the number of samples $N_{S\text{-red}}$, the higher the number of possible iterations and consequently the estimation precision increases.

Another possible approach is to apply the algorithm to a few DFTs, in general five or six, and to find peaks in the individual spectra. A function fitting these points is achieved by regression analysis. The function maximum, found by derivation, indicates the frequency on the abscissa and the amplitude at the ordinate. A zoom of the overlapped spectrum peak is shown in Figure 19, highlighting the points employed in the regression analysis.

4.3. Numerical comparison of the presented methods

The tests are performed employing the harmonic signal with the frequency $f = 5\,\text{Hz}$ and observation time length $T_S = 1.1\,\text{s}$. It contains $N_S = 11000$ samples generated with the sampling rate $F_S = 10000\,\text{Hz}$. To find out how efficient the interpolation methods are, 11 scenarios regarding the analysis time length were created. Therefore, a sequence of $N_{S\text{-it}} = 200$ samples was iteratively removed from the original signal, which corresponds to a step $T_{S\text{-it}} = 0.2\,\text{s}$. A DFT was applied to each resulting signal. Table 6 indicates the signal time lengths, the frequency resolution and the three

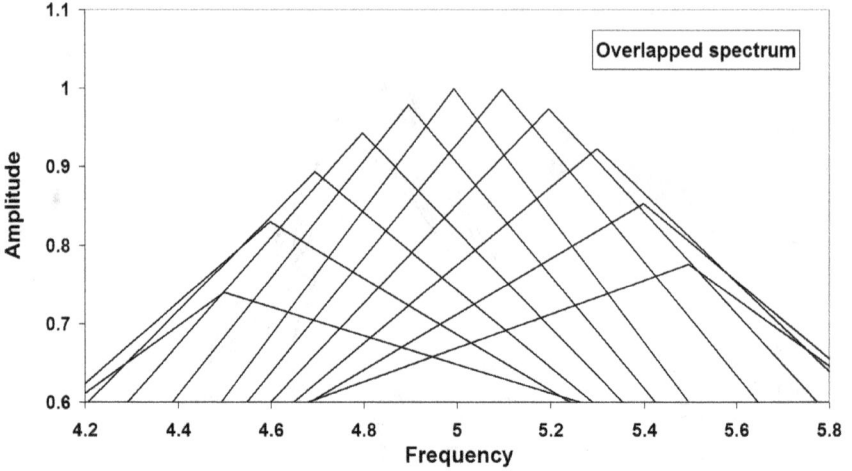

Figure 19. Detailed view of the overlapped spectrum peak.

Table 6. The coordinates of the maximizer and its two neighbors for the 11 observation times.

T_S (s)	Δf (Hz)	f_{j-1} (Hz)	A_{j-1} (mm/s^2)	f_j (Hz)	A_j (mm/s^2)	f_{j+1} (Hz)	A_{j+1} (mm/s^2)
1.1	0.909091	3.636364	0.245713	4.545455	0.666935	5.454545	0.608941
1.08	0.925926	3.703704	0.242976	4.62963	0.780512	5.555556	0.483371
1.06	0.943396	3.773585	0.20824	4.716981	0.866372	5.660377	0.36156
1.04	0.961538	3.846154	0.150813	4.807692	0.92993	5.769231	0.239611
1.02	0.980392	3.921569	0.080904	4.901961	0.975751	5.882353	0.116593
1	1	4	0	5	1	6	0
0.98	1.020408	4.081633	0.100583	5.102041	0.991669	6.122449	0.096852
0.96	1.041667	4.166667	0.228265	5.208333	0.941507	6.25	0.162058
0.94	1.06383	4.255319	0.378165	5.319149	0.850485	6.382979	0.191949
0.92	1.086957	4.347826	0.533492	5.434783	0.73149	6.521739	0.193801
0.9	1.111111	4.444444	0.674068	5.555556	0.603114	6.666667	0.181891

relevant amplitudes and related frequencies. The procedure was repeated for the signals windowed with the Hamming function in order to fulfill the conditions imposed by the Grandke method. To save space, these results are not explicitly presented but the image of peak amplitudes will be later presented.

The ratios α and the subsequent frequency correction δ terms were calculated by employing the relations presented in Section 4.1. The results are given in Table 7, for the methods supported by two points, and in Table 8 for those supported by three points. Also, the corrected frequencies

Table 7.   The correction terms and the values of the corrected frequencies for the methods considering two points for interpolation.

T_S (s)	Grandke		Quinn		Jain	
	$\delta(-)$	f_{corr} (Hz)	$\delta(-)$	f_{corr} (Hz)	$\delta(-)$	f_{corr} (Hz)
1.1	0.431819	5.608941	0.583334	5.075758	0.730769	4.3007
1.08	0.147347	5.743285	0.452018	5.048165	0.7626	4.409815
1.06	−0.11666	5.667997	0.316411	5.015481	0.806218	4.534167
1.04	−0.38537	5.588791	0.193569	4.993814	0.860454	4.673512
1.02	−0.67979	5.507212	0.090411	4.990598	0.923434	4.826895
1	−1	5.425926	0	5	0	5
0.98	0.471654	5.348458	0.112877	4.991594	−0.10824	4.991594
0.96	0.245568	5.278022	0.320039	4.991758	−0.20791	4.991758
0.94	0.010056	5.216292	0.800654	5.009067	−0.29148	5.009067
0.92	−0.20059	5.163114	2.694431	5.04301	−0.36043	5.04301
0.9	0.416667	4.812744	0.607143	5.119047	−0.43182	5.075759

Table 8.   The correction terms and the values of the corrected frequencies for the methods considering three points for interpolation.

T_S (s)	Jacobsen		Ding		Voglewede	
	$\delta(-)$	f_{corr} (Hz)	$\delta(-)$	f_{corr} (Hz)	$\delta(-)$	f_{corr} (Hz)
1.1	0.75796	5.30341	0.23871	4.78417	0.37898	4.92443
1.08	0.28801	4.91764	0.15953	4.78916	0.14400	4.77363
1.06	0.13183	4.84881	0.10675	4.82373	0.06591	4.7829
1.04	0.06043	4.86812	0.06725	4.87494	0.03021	4.83790
1.02	0.02034	4.92230	0.03041	4.93238	0.01017	4.91213
1	0	5	0	5	0	5
0.98	−0.00209	5.09995	−0.0031	5.09890	−0.0010	5.10099
0.96	−0.04435	5.16397	−0.0497	5.15862	−0.0221	5.18615
0.94	−0.16467	5.15448	−0.1310	5.18806	−0.0823	5.23681
0.92	−0.46173	4.97305	−0.2328	5.20192	−0.2308	5.20391
0.9	−1.40514	4.15041	−0.3373	5.21823	−0.7025	4.85298

are indicated in these tables. The frequency evolution with the analysis time is graphically represented in Figures 20–22 in order to facilitate comparison.

The maximum possible error induced by the standard frequency evaluation method is 0.5 Hz for an analysis time of 1 s. From Figures 20–22, one can observe that interpolation methods improve the frequency readability, all analyzed methods providing results with lower errors. The worst results are obtained, as expected, for the signals with the analysis time 0.9 s and 1.1 s,

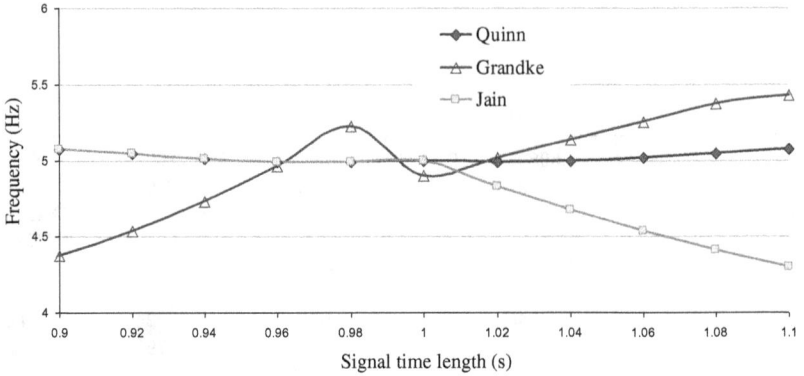

Figure 20.   Accuracy of the frequency estimation methods based on the analysis of two spectral lines.

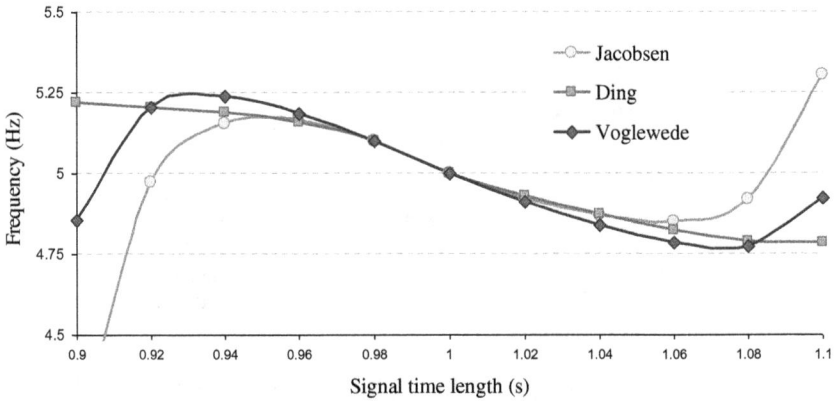

Figure 21.   Accuracy of the frequency estimation methods based on the analysis of three spectral lines.

respectively. These are particular time lengths for the 5 Hz sinusoid, where $T_S = (n + 0.5)T$ and two similar amplitudes are observed in the spectrum.

A second observation concerns the error evolution. It is periodic and has the same period T as the harmonic frequency component. Generally, the error is lowest if the signal contains an entire number of cycles, i.e. $T_S = nT$ and is most important for $T_S = (n + 0.5)T$.

Regarding the error range, the method proposed by Quinn is the most reliable. Good results are achieved also with the methods developed by Voglewede and Ding. Some other methods estimate well the frequencies at specific intervals.

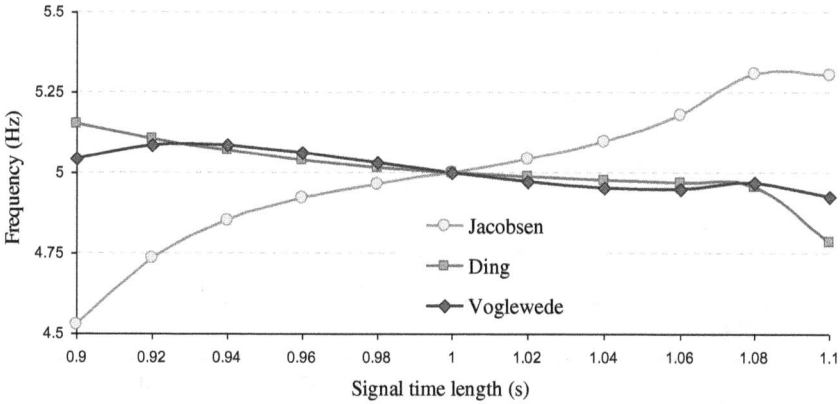

Figure 22. Precision of frequency estimation methods based on the analysis of three spectral lines — a signal preprocessing by windowing was applied.

Windowing the signal before applying the DFT improves the estimated frequency results achieved by employing the three-point interpolation methods. The improvement can be observed by comparing Figure 21 with 22.

The final observation is regarding the evaluator proposed by Çandan. It takes effect just for signals containing less than 150 samples, which is not usual for vibration measurements. For higher number of samples, it indicates the same values as the original Jacobsen method.

A strategy to guarantee that all three points used for interpolation belong to the main lobe is to take them from different spectra obtained by signal truncation. The evaluation algorithm presented in Section 4.2 makes use of this property.

Implementation of the algorithm is made for data presented in Table 7. The data result is achieved by iteratively truncating the original signal that has the length $T_{S\text{-prim}} = 1.1$ s and contains $N_{S\text{-prim}} = 11000$ samples with $N_{S\text{-it}} = 200$ samples by iteration, i.e. $T_{S\text{-it}} = 0.2$ s. The number of iterations is $\mu = 10$ and the final sequence contains $N_{S\text{-fin}} = 9000$ samples for the time length $T_{S\text{-fin}} = 0.9$ s. The peaks of the individual spectra are reproduced in Figure 23. Supplementary peak values for the DFTs extracted from the tapered (with a Hamming window) signals are shown. Note that, in general, fewer iterations are requested. The 10 iterations employed here are for demonstration purposes only.

Six cases were selected for simulation in order to cover possible specific situations. The coordinates of the chosen points are indicated in Table 9. For the first three cases, the points are taken from DFTs extracted from the

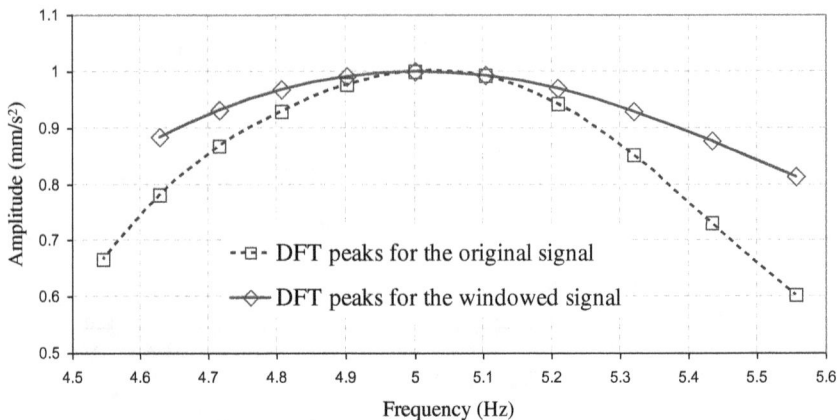

Figure 23.   Peaks achieved in the individual DFT spectra centralized in the overlapped spectrum.

Table 9.   Coordinates of the points employed in the six simulation cases and the resulting corrected frequency.

Case no.	f_{j-1} (Hz)	A_{j-1} (mm/s²)	f_j (Hz)	A_j (mm/s²)	f_{j+1} (Hz)	A_{j+1} (mm/s²)	f_{corr}
1	4.807692	0.92993	5.102041	0.991669	5.434783	0.73149	5.02118
2	4.807692	0.92993	5.208333	0.941507	5.555556	0.603114	5.01863
3	4.716981	0.866372	5.208333	0.941507	5.555556	0.603114	5.01971
4	4.807692	0.967896	5.102041	0.993213	5.434783	0.876081	5.01653
5	4.807692	0.967896	5.208333	0.969509	5.555556	0.812744	5.01116
6	4.716981	0.93179	5.208333	0.969509	5.555556	0.812744	5.02389

signals in the absence of tapering, whereas for the next three cases a window was first applied to the signals. Functions fitting the groups of points were contrived by regression analysis and the abscissa of their maxima found by derivation. The achieved values represent the estimated frequency, indicated in the last column of Table 9. The regression functions for cases 1 and 4 are presented in Figure 24 in graphical form and as mathematical relations.

Taking a look at Table 9, one can observe that the precision ensured by involving this algorithm is considerably improved, all errors being less than 0.5%. This error limit is valid for any combinations of time lengths, but note that the amplitudes should belong to the same main lobe. The condition is fulfilled if the time limits are set in accordance with $T_{S\text{-prim}} = (n + 0.45)T_{E\text{-prim}}$ and $T_{S\text{-fin}} = (n - 0.45)T_{E\text{-prim}}$, where the period

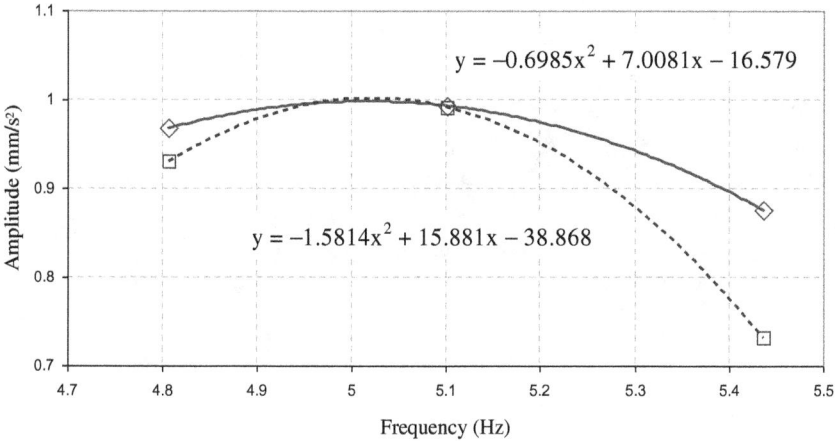

Figure 24.   Regression functions for the cases 1 and 4.

$T_{E\text{-prim}} = 1/f_{E\text{-prim}}$ results from the coarse frequency estimation. Regarding the frequency estimation improvement, it was shown that the algorithm is insensitive to windowing.

Comparing the methods to improve frequency estimation based on interpolation clearly shows that the most reliable one involves the overlapped spectrum with the narrow error range recommending this method for early observation of structural changes with application in damage detection.

5. Tests Performed on Real Signals

Since the method should be appropriate for damage detection, it has to work well in real conditions. The experiment is therefore made on a real beam, made of steel, having length $L = 0.89$ m, width $B = 0.05$ m and thickness $H = 0.05$ m, subsequently, the cross-sectional area $A = 250 \cdot 10^{-6}$ m^2 and the moment of inertia $I = 520.833 \cdot 10^{-12}$ m^4.

Figure 25 illustrates the experimental setup. The beam is fixed at one end and has the other end free. The measurement consists of a laptop, an NI cDAQ-9171 compact chassis with an NI 9234 four-channel dynamic signal acquisition module and a Kistler 8772 accelerometer. The algorithm was implemented in LabVIEW by creating two VIs, one for acquisition and storage and the second for signal analysis.

Damage detection methods developed by the authors imply measuring several weak-axis bending vibration modes.[27–30] With this example, we

Figure 25.   View of the experimental stand.

aim to demonstrate that the frequency components can be estimated with precision, and identification of small frequency changes is possible. We also show that the effect of leakage is not extended to the neighbor frequency components.

A first measurement was made for the intact beam. The sampling frequency was set to $F_S = 25{,}000$ Hz in order to allow a large number of iterations. An observation time $T_S = 2$ s was selected, requiring $N_S = 50000$ samples in the discrete signal. For the first mode, the DFT indicated the coarse estimated frequency $f_{E\text{-prim}} = 5$ Hz, thus this mode has the estimated period $T_{E\text{-prim}} = 0.2$ s. Setting $n = 8$ cycles and $\mu = 50$ iterations, the algorithm implemented in the VI calculates all necessary parameters and returns the peak amplitude and estimated frequency. Since dense spectral lines are ensured in the overlapped spectrum, the indicated frequency can be considered as the corrected one.

The main lobe formed around the indicated frequency, which is 5.03999 Hz, is presented in Figure 26. One can also observe on the right bottom graph that leakage is not extended to other modes. Since each frequency component must be separately analyzed, an appropriate excitation can be applied in order to increase the amplitude of the considered vibration mode. This way, the effect of leakage emanated from other vibration modes on the analyzed component is negligible.

For a reduced number of iterations, $\mu = 5$, the result is an overlapped spectrum indicating six frequency-amplitude pairs, which are shown in Figure 27. The maximizer indicated here cannot be considered as the corrected frequency, but he and his two neighbors can be involved in regression analysis. In Figure 28, the six amplitudes are taken for analysis,

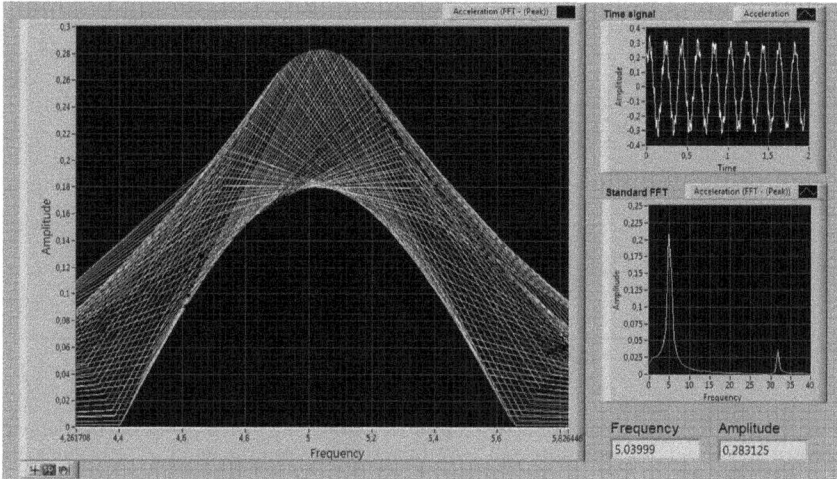

Figure 26. Dense overlapped spectrum. The local peak amplitude and the related frequency are automatically indicated by the VI.

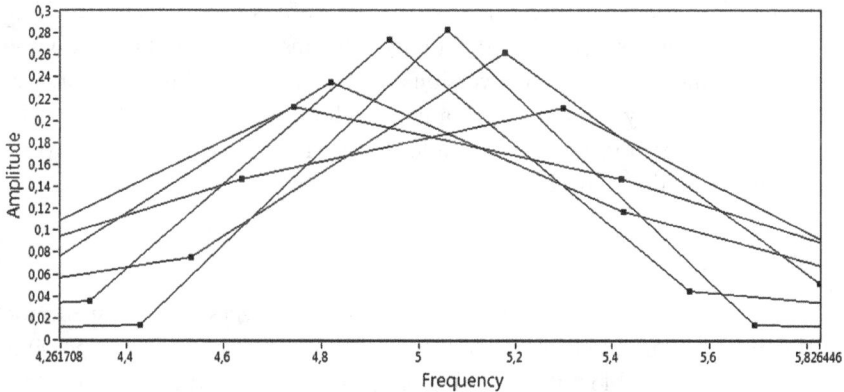

Figure 27. Overlapped spectrum for the intact beam achieved for five iterations. The frequency-amplitude pairs are automatically extracted by the VI.

the continuous line and the upper relation staying for this case. From here results the corrected frequency $f_{\text{corr}} = 5.03748\,\text{Hz}$.

At the distance $x_D = 0.19\,\text{m}$ from the fixed end, a saw cut damage with depth 1 mm and width 2 mm is produced and the measurements are repeated. Now, the corrected frequencies are $f_{\text{corr}} = 4.88993\,\text{Hz}$ for the dense spectrum, and $f_{\text{corr}} = 4.90829\,\text{Hz}$ if the number of iterations is small.

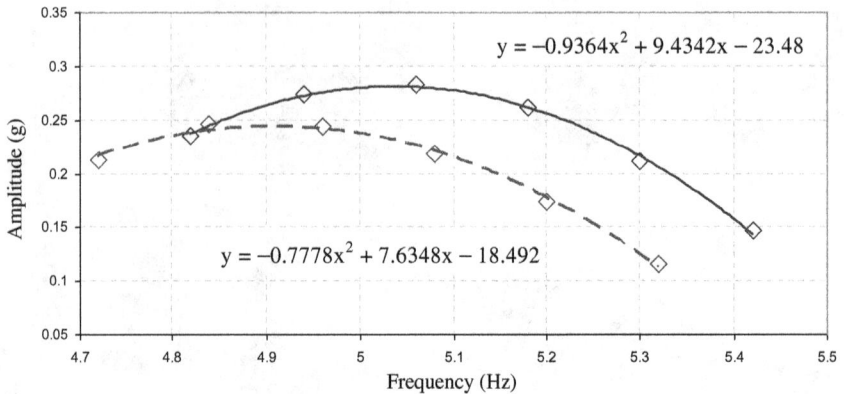

Figure 28.   Regression analysis based on the amplitudes extracted from the overlapped spectrum with six relevant spectral lines: intact beam (continuous line) and damaged beam (dashed line).

The regression function for the damaged beam is presented in Figure 28 with a dashed line and its mathematical relation is presented below. One observes that the relatively small frequency shift is easily observed. Note that for both cases, intact and damaged beams, the difference between results obtained from the two versions of the method is insignificant, less than the frequency shift due to damage. The high precision achieved by involving this estimation method allows reading frequencies with several digits after the decimal point.

6. Conclusion

Damage detection by frequency shift evaluation requests precise estimation of the measured frequencies. Since the acquisition time is short, the possible errors are comparable with the frequency shifts due to damage, therefore observing the occurrence of damage in the early stage is impossible. Actual simple methods as zero-padding, signal windowing or interpolation do not offer a real frequency readability improvement. From the numerical simulation performed, we found out that the errors depend on the acquisition time and can achieve 10%.

An alternative to the actual interpolation methods is considering the frequency-amplitude pairs captured from different spectra of the same signal. Overlapping these spectra the distance between spectral bins dramatically decreases and makes possible the reading of frequencies with high precision.

The proposed method was tested on generated signals with known frequency and on signals acquired from healthy and damaged structures. Accurate frequency identification was accomplished in all cases. The damage, which is observable for an acquisition time above 5 s if standard frequency evaluation is employed, was easily identified. Applying this method becomes advantageous especially for higher vibration modes, where rapid decay limits the length of the signal.

References

1. Doebling, S. W., Farrar C. R., and Prime, M. B. A summary review of vibration based damage identification methods. *Shock and Vibration Digest* **30**(2), pp. 91–105 (1998).
2. Salawu, O. S. Detection of structural damage through changes in frequency: A review. *Engineering Structures* **19**(9), pp. 18–723 (1997).
3. Morassi, A. and Vestroni, F. *Dynamic Methods for Damage Detection in Structures*, Vol. 499, CISM Courses and Lectures, Springer (2008).
4. Gillich, G. R. and Praisach, Z. I. Modal identification and damage detection in beam-like structures using the power spectrum and time-frequency analysis. *Signal Process* **96**(Part A), pp. 29–44 (2014).
5. Friswell, M. I. Damage identification using inverse methods. *Philosophical Transactions of the Royal Society A* **365**, pp. 393–410 (2007).
6. Gillich, G. R., Praisach, Z. I., Abdel Wahab, M., and Vasile, O. Localization of transversal cracks in sandwich beams and evaluation of their severity. *Shock and Vibration* **2014**, 607125, pp. 10 (2014).
7. Rytter, A. *Vibration Based Inspection of Civil Engineering Structures*. Ph.D. Thesis, Aalborg University, Denmark (1993).
8. Almeida, R. Urgueira, A., and Maia, N. M. M. Further developments on the estimation of rigid body properties from experimental data. *Mechanical System and Signal Processing* **24**(5), pp. 1391–1408 (2010).
9. Hutin, C. Modal analysis using appropriated excitation techniques. *Sound Vibrations* **34**(10), pp. 18–25 (2000).
10. Sakaris, C., Sakellariou J. S., and Fassois, S. How many vibration response sensors for damage detection & localization on a structural topology? An experimental exploratory study. *Key Engineering Materials* **569–570**, pp. 791–798 (2013).
11. Gillich, G. R., Praisach, Z. I., and Iavornic, C. M. Reliable method to detect and assess damages in beams based on frequency changes. In *Proceeding of the ASME International Design Engineering Technical Conferences and Computers and Information in Engineering Conference (IDETC/CIE 2012)*, Vol. 1, pp. 129–137, Chicago, USA (Aug, 2012).
12. National Instruments. *LabVIEW Analysis Concepts*, Part Number 370192C-01 (Mar, 2004).
13. Minda, A. A., Gillich, N., Mituletu, I. C., Ntakpe, J. L., Manescu, T., and Negru, I. Accurate frequency evaluation of vibration signals by

multi-windowing analysis. *Applied Mechanics and Materials* **801**, pp. 328–332 (2015).

14. Chioncel, C. P., Gillich, N., Tirian, G. O., and Ntakpe, J. L. Limits of the discrete fourier transform in exact identifying of the vibrations frequency. *Romanian Journal of Acoustics and Vibration* **12**(1), pp. 16–19 (2015).

15. Donciu, C. and Temneanu, M. An alternative method to zero-padded DFT. *Measurement* **70**, pp. 14–20 (2015).

16. Andria, G., Savino, M., and Trotta, A. Windows and interpolation algorithms to improve electrical measurement accuracy. *IEEE Transactions on Instrumentation and Measurements* **38**(4), pp. 856–863 (1989).

17. Abed, S. T., Dallalbashi, Z. E., and Taha, F. A. Studying the effect of window type on power spectrum based on MatLab. *Tikrit Journal of Engineering Science* **19**(2), pp. 63–70 (2012).

18. Djukanović, S., Popović, T., and Mitrović, A. Precise sinusoid frequency estimation based on parabolic interpolation. In *Proceeding of the 24th Telecommunications Forum TELFOR*, pp. 1–4, Belgrade, Serbia (2016).

19. Grandke, T. Interpolation algorithms for discrete fourier transforms of weighted signals. *IEEE Transactions on Instrumentation and Measurements* **32**, pp. 350–355 (1983).

20. Quinn, B. G. Estimating frequency by interpolation using fourier coefficients. *IEEE Transactions on Signal Processing* **42**, pp. 1264–1268 (1994).

21. Jain, V. K., Collins, W. L., and Davis, D. C. High-accuracy analog measurements via interpolated FFT. *IEEE Transactions on Instrumentation and Measurements* **28**, pp. 113–122 (1979).

22. Ding, K., Zheng, C., and Yang, Z. Frequency estimation accuracy analysis and improvement of energy barycenter correction method for discrete spectrum. *Journal of Mechanical Engineering* **46**(5), pp. 43–48 (2010).

23. Voglewede, P. Parabola approximation for peak determination. *Global DSP Magazine* **3**(5), pp. 13–17 (2004).

24. Jacobsen, E. and Kootsookos, P. Fast, accurate frequency estimators. *IEEE Signal Processing Magazine* **24**(3), pp. 123–125 (2007).

25. Çandan, C. A method for fine resolution frequency estimation from three DFT samples. *IEEE Signal Processing Letters* **18**(6), pp. 351–354 (2011).

26. Mituletu, I. C., Gillich, N., Nitescu, C. N., and Chioncel, C. P. A multi-resolution based method to precise identify the natural frequencies of beams with application in damage detection. *Journal of Physics: Conference Series* **628**(1), 012020 (2015).

27. Gillich, G. R. and Praisach, Z. I. Detection and quantitative Assessment of Damages in Beam Structures Using Frequency and Stiffness Changes. *Key Engineering Materials* **569**, pp. 1013–1020 (2013).

28. Gillich, G. R. and Praisach, Z. I., Robust method to identify damages in beams based on frequency shift analysis. In *Proceeding SPIE 8348, Health Monitoring of Structural and Biological Systems 2012*, 83481D, San Diego, USA (Mar, 2012).

29. Gillich, G. R., Maia, N. M. M., Mituletu, I. C., Praisach, Z. I., Tufoi, M., and Negru, I. Early structural damage assessment by using an improved frequency evaluation algorithm. *Latin American Journal of Solids and Structures* **12**(12), pp. 2311–2329 (2015).
30. Gillich, G. R., Praisach, Z. I., and Negru, I. Damages influence on dynamic behaviour of composite structures reinforced with continuous fibers. *Materials and Plastics* **49**(3), pp. 186–191 (2012).

Chapter 5

Damage Localization Based on Modal Response Measured with Shearography

J. V. Araújo dos Santos[*,‡] and H. Lopes[†,§]

*IDMEC, Instituto Superior Técnico, Universidade de Lisboa
Av. Rovisco Pais, 1049-001 Lisboa, Portugal
†DEM, ISEP, Instituto Politécnico do Porto
Rua Dr. António Bernardino de Almeida
431, 4249-015 Porto, Portugal
‡viriato@tecnico.ulisboa.pt
§hml@isep.ipp.pt

This chapter describes the application of shearography, which is a non-contact, full-field, and high-resolution optical method, for the measurement of modal response of beams and plates and subsequent damage localization. A review of literature on shearography and related interferometric techniques for vibration analysis is also presented. Since shearography is based on speckle interferometry, the physical and mathematical bases of speckle phenomenon and wave interference are described. The main techniques available to evaluate the phase maps obtained from the speckle patterns are considered and illustrated. Different processes of filtering and unwrapping the phase maps, which must be carried out in order to obtain the gradients of the modal displacement fields and thus the modal rotation field, are also presented in this chapter. Case studies of damage localization using measurements of modal response with two distinct shearography systems are also reported and discussed. Both single and multiple damages in aluminum beams are created using a saw or a milling machine. In order to test the accuracy and effectiveness of the damage localization approach, several amounts of damages are considered. It is found out that very small damage can be localized using higher order derivatives of the mode shapes, namely the fourth order.

Keywords: Damage localization; Modal response; Shearography; Speckle interferometry; Modal rotation field; Contactless measurement; Electronic speckle pattern interferometry (ESPI); Modal curvature field; Phase maps; Temporal phase shifting.

1. Introduction

A considerable amount of research in recent years has been focused on the improvements of well-established damage localization methods based on vibration data, such as the curvature mode shape method, proposed by Pandey *et al.*[1] and similar techniques.[2-4] Most improvements are achieved by applying or incorporating more complex approximations of the derivatives of the response data. For example, Sazonov and Klinkhachorn[5] developed a method with an optimal sampling interval in the discretization of the mode shapes of beams. Guan and Karbhari[6] proposed the differentiation of sparse and noise measurements with a fourth-order central difference method instead of the second-order method. The uncertainty in geometry and measurements can also, according to Chandrashekhar and Ganguli,[7] be alleviated by a fuzzy logic approach, together with a damage indicator based on curvatures. In order to reduce the influence of the measurement noise, Rucevskis and Wesolowski[8] proposed a damage indicator based on the average sum of the mode shape curvature squares for all the modes of interest. A damage indicator combining deterministic and stochastic approaches is proposed by Tomaszewska.[9] Radzieński[10] found out that wavelet transforms lead to improvements, being most effective, noise independent, and versatile. Solis *et al.*[11] also used a method based on wavelet transforms. This method includes a curve fitting approach, which also works as a smoothing process. The Teager energy operator wavelet transform is applied by Cao *et al.*[12] to mode shapes of beams. By taking into account the width of the derivative interval and the resolution of the resulting derivative, Xu *et al.*[13] proposed a modification of the second-order central finite difference. The windowed Fourier ridges algorithm is introduced by Yang *et al.*[14] to extract the curvature of the mode shape pre-treated with a numerical interpolation. Nevertheless, besides the improvements coming from the application of more efficient numerical techniques, one may also use more reliable and efficient experimental techniques to measure the modal response than those commonly used. For instance, if we use conventional experimental modal analysis techniques with force transducers and accelerometers as actuators and sensors, which are attached to the vibrating structure, their influence may change significantly their dynamic behavior. In order to minimize this problem, we can adopt techniques which excite and sense the structure in a contactless manner.

Among several non-contact measurement techniques available, optical methods are the most sensitive, and their use has dramatically increased in recent years.[15] Examples of applications of optical methods in the measurement of modal response for damage localization can be found in Refs. 16–19. The modal response measured in these works is the modal displacement fields, which describe the mode shape. Because damage localization based on curvatures needs second-order derivatives of these fields, an alternative to the techniques used in the previous works is shearography, also known as shearing or shear interferometry. This technique measures directly the displacement gradients in a selected direction and can be seen as a differentiation procedure achieved by optical means. Shearography relies on the speckle phenomenon, which appears when a rough and diffuse surface is illuminated by coherent light. In most applications of shearography, the coherent light used is one or more laser beams. The origins of shearography can be traced back to developments in speckle interferometry made in the early 1970s.[20–22] Speckle interferometry is closely related to holographic interferometry. The speckle was initially seen as noise in the reconstructed image obtained by holography and, thus, an undesired effect.[23] Later, it was found out that the speckle contains information about the surfaces. A historical description of the study of the speckle phenomena can be found in Refs. 24–26, and a summary of the first progresses on speckle interferometry reported in the literature are presented in Ref. 27.

This chapter begins with a review of the literature reporting measurements of vibrating objects by shearography and related speckle interferometric techniques. Afterwards, the principles of speckle interferometry and its relation with shearography are described. Particular attention is paid to the different techniques available to evaluate the phase maps from the speckle patterns. Methods of filtering and unwrapping the phase maps in order to determine the continuous displacement field gradients are also presented. Finally, case studies of damage localization using measurements of modal response with shearography are presented and discussed. These studies show that the sensitivity of the measurements is such that they allow the localization of small and multiple damages.

2. Review of Shearography for Vibration Analysis

Although the first works on the use of the speckle effect for measurement of static displacements and contour mapping appeared in 1968, according to Cloud,[27] quickly the interest on the application of this effect also extended to vibrating bodies in the next and subsequent years. Archbold *et al.*[28] built an instrument based on the fact that when a surface is vibrating, the nodal areas can be visible because in these regions the speckles have a higher contrast than in other areas. They compared the visually observed regions with high contrast speckles to recorded holograms of a vibrating square plate. Despite the fact that the method only determines the nodal areas, being thus essentially an observational method, it does not involve the photographic process and does not need the drift stability which is essential in recording the hologram. The experimental arrangement of Archbold et al. was also used to detect the amplitude of vibration by Ek and Molin.[29] Tiziani[30] was able to apply the speckle phenomenon in the computation of the amplitude of the mechanical oscillations of a tuning fork of an electronic watch. Similar conclusions on the use of the speckle effect for vibration analysis were found by Fernelius and Tome,[31] who studied a tin-clad-steel glovebox port cover acoustically excited at different frequencies. Since the analyzed surface was planar, it was possible to compare the speckle patterns to Chladni strewn sand patterns and considerable similarities between them were observed. A clamped square steel plate was also studied to show that the effect could be repeated and observed in different structures. Finally, an aluminum beer can, which presents a curved surface, was also studied and modes related to standing waves around the circumference of the can could be identified. It is worth mentioning that among the conclusions made by Fernelius and Tome,[31] we can find one regarding the advantages of the proposed approach of being a non-contact method and that no mass is added to the vibrating object, therefore not changing the resonance frequencies. Typical vibration modes of a circular disc, excited with a crystal, were obtained directly from a television monitor by Butters and Leendertz.[23] Thus, in their method, it is possible to watch the changes in the mode patterns with changing frequency, because a real-time processing is involved, which, unlike stroboscopic real-time interferometry, does not need synchronization of the strobe frequency and there are no switching frequency limits. The aforementioned works are, among others also reporting static measurements, the first ones employing optical configurations and other instruments which gave origin to what is nowadays called electronic speckle pattern interferometry (ESPI) or TV-holography.

An alternative to ESPI, which allows the visualization of patterns of the mode-shape slopes or derivatives of vibration amplitudes, instead of the mode shapes or the vibration amplitudes themselves, is accomplished by shearography. Contrary to ESPI, in shearography the interferometer is less sensitive to external perturbations, such as convective currents or vibrations. Thus, it is possible to use this method outside the controlled ambient of a laboratory. A comprehensive description of the differences between ESPI and shearography can be found in Ref. 32. One of the first works reporting measurements of vibration characteristics using shearography was published by Hung and Taylor.[22] A rectangular plate, fully clamped, is analyzed at two distinct frequencies and fringe patterns showing slopes, i.e. first-order derivatives in the x direction, of the modal amplitudes are presented. A fringe pattern of slopes of modal amplitudes of a plate vibrating at its fundamental mode can also be found in Ref. 33, which reports improvements in the speckle shearing interferometric method presented in Ref. 22. A different optical configuration to obtain partial slope contour fringes is proposed by Chiang and Juang,[34] who reported partial anti-nodal and partial slope contour fringes of a square plate vibrating at various frequencies. Their method is also applied to a vibrating sector of a thin circular cylindrical shell. Although this method allows the derivative direction and the sensitivity to be varied after recording, it produces fringes of poor quality and is less tolerant to rigid body rotations.[35]

Since these seminal works, and related ones reporting the use of shearography for vibration analysis,[36–38] a large improvement in digital recording and image processing techniques has taken place, allowing better quality in the visualization of surface vibrations. Presently, the digital recording media rely usually in video sensors, which avoid the use of consumables needed in the photographic and thermoplastic versions of shearography, thus allowing real-time measurements.[39] Also of particular importance is the application of digital image processing techniques to the acquired speckle pattern. According to Nakadate *et al.*[40] it is possible to improve the quality of the fringe patterns and to calculate the surface strain components by digitally processing the speckle pattern by a computer. Besides presenting the fringe patterns contouring of the horizontal slope of the displacement of a circular plate loaded at the center, Ref. 40 also presents the horizontal slope of the normal vibration amplitude of this plate. Another example of application of the digital version of shearography to modal analysis can be found in Ref. 41. In this work, Ng and Chau[41] obtained the modal slope shapes of the first five modes of a thin, fully clamped square plate. It was found out that the shearing speckle modal slope shapes, which are viewed at video frame-rate, are consistent with theoretical derivations. Yang *et al.*[42] also explored the possibilities of digital shearography for vibration analysis. They developed a method combining continuous and stroboscopic illuminations. The first type of illumination is well suited for qualitative vibration analysis and non-destructive testing by dynamic excitation. If one wants further evaluation of the shearogram in a certain frequency, the stroboscopic illumination can be selected by adjusting a controller and an accurate phase map can be obtained. A rectangular steel plate fully clamped with four grooves on the backside was excited at various frequencies and observed in real-time using continuous illumination, showing that all the grooves can be displayed and thus proving the usefulness of digital shearography as a qualitative non-destructive tool. A turbine blade clamped at one end and excited harmonically with a piezoelectric crystal on the other end was also analyzed by Yang *et al.*[42] Phase maps, unwrapped phase distributions and the 3D plot of the displacement derivative field are presented. The displacement field and the flexural strain field are obtained, respectively, by integration and differentiation of the unwrapped phase map. The measurement of modal damping of a cantilever beam excited by a mini vibration shaker using shearography was carried out by Wong and Chan.[43] The proposed method has been validated by using an accelerometer-based modal analysis method and it was found out that the difference is about 0.7%. A cylinder excited harmonically by an electrodynamic shaker was analyzed by Casillas *et al.*[44] The small in-plane and out-of-plane vibration amplitudes are estimated more accurately by the proposed phase recovery technique in the stroboscopic shearography. The experimental results compare well with those from a finite element simulation. A dual function system that integrates digital speckle pattern interferometry and digital shearography into a single system to visualize mode shapes at fringe pattern level was proposed by Bhadury *et al.*[45] The system was used to investigate vibration mode shapes of an aluminum plate clamped at the bottom and a piezoelectrically actuated valveless micro-pump, vibrating sinusoidally at different resonant frequencies.

Most of the aforementioned works only describe techniques for the measurements of vibration amplitudes. Nevertheless, it is possible to determine also the temporal characteristics of the vibration cycle and the phase with special optical configurations and post-processing of images. For instance, Valera and Jones[46] proposed a technique to determine the vibration phase and the derivative sign. They used a fiber-based speckle shearing interferometer with sinusoidal phase modulation between the interfering images to out-of-plane time-averaged vibration analysis. They applied it to a circular aluminum plate of 14 cm of diameter, excited at two different frequencies. Phase distributions for any state of the vibration cycle can also be obtained by a method proposed by Somers *et al.*[47] thus allowing both temporal and spatial characterization of harmonic vibrations. An aluminum plate of 240 × 240 mm, with a thickness of 0.5 mm, elastically supported by four springs, attached to the corner of the plate and excited by a shaker at one of the resonance frequencies of the plate was analyzed by the proposed method. The results show the reversal of phase for first and second halves of the vibration cycle. By accumulating phase differences for all or a selection of measurement intervals, it is possible to directly visualize unwrapped positive and negative parts of the vibration cycle.

Because shearography measures the gradient of the displacements, the strains can also be directly obtained. If the measured strains present some perturbations, usually in terms of strain concentration, one can promptly correlate these perturbations with flaws or defects in the material. This is one of the reasons why shearography has been used for non-destructive testing since it was first developed. Along with some works cited before, we can find more applications of digital shearography to non-destructive testing of composite materials using dynamic loading in Refs. 48 and 49. Recently, the post-processing of the unwrapped phase map to obtain the modal curvature field has also been used to localize damage in isotropic[50] and composite materials.[51–53]

Besides the measurement of structures undergoing vibrations, shearography is also capable of measuring the displacement of gradients of structures thermally loaded or subjected to pressure or vacuum, among other kinds of excitations. Comprehensive state-of-the-art descriptions and reviews of shearography and its applications to several types of measurements have been published,[54,55] showing extensive improvements over the years on the various developed techniques. In addition to works describing different types of experimental setups and methods for post-processing of images, Refs. 54 and 55 also refer to the application of shearography to transient vibrations. In this case, a high-speed camera or a pulsed laser are used to investigate the time-dependent displacement derivative.[54] Works reporting comparisons of several techniques to shearography for non-destructive testing are also worth mentioning.[56–59]

3. Principles of Speckle Interferometry and Shearography

The principles of speckle interferometry, namely the speckle characteristics and the interference phenomenon, and its relation with shearography are described in this section. Techniques to evaluate the phase maps, which are based on the post-processing of recorded speckle patterns, are also presented. Several methods of filtering and unwrapping the phase maps in order to determine the continuous displacement field gradients are also presented.

3.1. *Speckle interferometry*

3.1.1. *Speckle characteristics*

Whenever a rough surface is illuminated by a light with proper characteristics, a speckle pattern is created due to the interference of multiple reflected spherical wave fronts. The wavelength of the light should be equal to or lower than the surface roughness. The light should also be coherent and that is why lasers are used in most applications. The speckles, which are represented in space as ellipsoids,[60] with dimensions varying with the distance from the reflecting surface, form a random pattern that, although stationary in time, is highly variable from point to point.[61] Therefore, the pattern presents a granular nature, as can be seen in Figure 1. There are two types of speckles: (1) objective and (2) subjective.[62,63] In the first case, there is no imaging system and the speckle dimension depends on the plane of observation and how the illumination of the surface is carried out. The subjective speckle is formed by the imaging system, and it is thus dependent on the limit of diffraction of this system. The image in the plane of observation is formed using a lens and an aperture. The control of the optical aperture allows adjustment of the speckle dimension, such that its diameter is adjusted to the size of the photoelectric sensor, as described in more detail in the following.

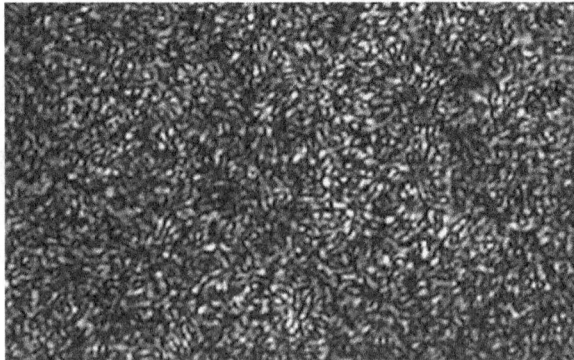

Figure 1. Speckle pattern obtained with an interferometer.

Each one of the points in an illuminated rough surface is, according to the Huygens's principle, a source of a spherical secondary wavelet. The electrical field in a given instant is given by[61]

$$E_n(r_n) = \frac{A_n}{r_n} e^{j(kr_n + \phi_n)} \quad \text{with } n = 1, 2, \ldots, N \tag{1}$$

where n denotes a specific point in the surface, N is the total number of points, A_n is the amplitude, r_n is the distance of the point to the plane of observation, k is the direction of the propagation and ϕ_n is the phase. All the source points contribute to the intensity of the incident light arriving at a point $P(x, y)$ in the observation plane and, therefore, we have[61]

$$E(x, y) = \sum_{n=1}^{N} \frac{A_n}{r_n} e^{j(kr_n + \phi_n)} \tag{2}$$

Equation (2) can be seen as representing the random walk problem in two dimensions and, thus, the central limit theorem can be applied, yielding, for the nth wave,

$$E_n(r_n) = \frac{|A_n|}{\sqrt{N}} e^{j\phi_n} \tag{3}$$

where $|A_n|/\sqrt{N}$ and ϕ_n are the amplitude and the phase, respectively. Because the surface has elementary areas randomly distributed, the amplitude and the phase of the speckle pattern are statistically independent. The phase follows a uniform distribution[61]:

$$P_\phi(\phi) = \begin{cases} \dfrac{1}{2\pi} & \text{if } -\pi \leq \phi < \pi \\ 0 & \text{if } \phi \geq \pi \quad \text{or } \phi < -\pi \end{cases} \tag{4}$$

whereas the intensity or amplitude follows a negative exponential probability distribution[61]:

$$P_I(I) = \begin{cases} \dfrac{1}{2\sigma^2} e^{-\frac{1}{2\sigma^2}} & \text{if } I > 0 \\ 0 & \text{if } I \leq 0 \end{cases} \tag{5}$$

where σ^2 is the variance of the joint probability function. If we consider that the mean value of the intensity is equal to $2\sigma^2$ and considering that the variance $2\sigma_I^2$ of the intensity is equal to the mean intensity, the contrast of the speckle pattern is always equal to unity.[61]

An important characteristic of the speckle is its dimension, which is directly related to the number of pixels of the CCD or CMOS sensors and hence the resolution of the measurement technique. In practice, the average dimension should be adjusted to the resolution of the acquisition system in order to resolve the speckle. It can be shown that the average dimension of the speckle can be obtained from the autocorrelation function of the intensity in the plane of observation. If we consider the Huygens–Fresnel principle and the diffraction limit of the imaging system, the autocorrelation is given by[61]

$$R(r) = \langle I \rangle^2 \left\{ 1 + \left| \frac{2J_1\left(\frac{\pi D_1 r}{\lambda z}\right)}{\frac{\pi D_1 r}{\lambda z}} \right|^2 \right\} \tag{6}$$

where $\langle I \rangle$ is the mean intensity, D_1 is the diameter of the lens pupil, J_1 is the Bessel function of the first kind of order one, λ is the wavelength, z is the distance from the observation plane to the lens pupil plane and $r = \sqrt{\Delta x^2 + \Delta y^2}$, Δx and Δy being the speckle dimensions in the x and y directions of the observation plane. The first minimum of the function $J_1(\pi D_1 r/\lambda z)$ gives us the average size d_s of the subjective speckle:

$$d_s = 1.22 \frac{\lambda z}{D_1} \tag{7}$$

Alternatively, we can define the average size as a function of the numerical aperture NA of the optical system and the wavelength λ:

$$d_s = 0.61 \frac{\lambda}{NA} \tag{8}$$

Equation (8) is valid for small apertures, since in this case, the numerical aperture is approximated by the ratio $D_1/2f$, where f is the focal distance, which is equal to z. A speckle pattern has a maximum spatial frequency given by the numerical aperture and the distance between the lens and the observation plane or by this distance, the wavelength, and the diameter of the lens pupil[64]:

$$f_{\max} = 2 \frac{D_1}{\lambda z} \tag{9}$$

The ideal characteristics of the optical system can be defined based on the above equations, allowing us to relate the speckle size with the dimension of each pixel in the CCD or CMOS sensor used to acquire the speckle intensities.

3.1.2. *Interference phenomenon*

The intensity of the light arriving at the photoelectric sensor is given by the flux of energy incident in an area per unit of time. In the case of a stationary wave, the intensity is given by

$$I(r) = \langle E(r,t)E^*(r,t) \rangle = \lim_{T_m \to +\infty} \frac{1}{T_m} \int_{-\frac{T_m}{2}}^{+\frac{T_m}{2}} E(r,t')E^*(r,t')dt' \quad (10)$$

where $*$ denotes the complex conjugate and

$$E(r,t) = A_0 e^{j(kr-\omega t+\phi)} \quad (11)$$

describes the space and time distribution of the light wave. In practice, the exposure time T_m is much larger than the period, $T_m \gg 2\pi/\omega$, such that, omitting the constants of proportionality, the intensity reduces to $I = |A_0|^2$. The superposition of two or more wave fronts of coherent light leads to the interference phenomenon. In the following, we consider two coherent plane wave fronts emitted from the same source, described by $E_1(r,t)$ and $E_2(r,t)$, with the same amplitude A_0 and frequency ω, but different traveling paths and phases ϕ_1 and ϕ_2. The result of the interference is a wave with amplitude dependent on the phase difference of the two initial wave fronts, such that[61]

$$(E_1 + E_2)(r,t)$$
$$= 2A_0 \cos\left[\frac{2\pi}{\lambda}\sin\left(\frac{\theta}{2}\right)r + \frac{\phi_1 - \phi_2}{2}\right] e^{j[\frac{2\pi}{\lambda}\cos(\frac{\theta}{2})r - \omega t + \frac{\phi_1+\phi_2}{2}]} \quad (12)$$

where θ is the angle formed by the vectors of the traveling paths of each initial wave and λ is the wavelength. From the definition of intensity in Equation (10), the interference intensity is given by

$$I(r) = (E_1 + E_2)(E_1 + E_2)^* = 4A_0^2 \cos^2\left[\frac{2\pi}{\lambda}\sin\left(\frac{\theta}{2}\right)r + \frac{\phi_1 - \phi_2}{2}\right]$$
$$(13)$$

The minimum and maximum values of the interference intensities occur, respectively, when

$$\frac{2\pi}{\lambda}\sin\left(\frac{\theta}{2}\right)r + \frac{\phi_1 - \phi_2}{2} = \frac{(2n+1)\pi}{2} \quad \text{for } n \in \mathbb{Z} \quad (14)$$

Figure 2.   Patterns obtained from the interference of two wave fronts.

and

$$\frac{2\pi}{\lambda} \sin\left(\frac{\theta}{2}\right) r + \frac{\phi_1 - \phi_2}{2} = n\pi \quad \text{for } n \in \mathbb{Z} \tag{15}$$

The first expression corresponds to destructive interference, when the two waves are in anti-phase, and the second corresponds to constructive interference, when the two waves are in phase. We call interference pattern to the intensities of a set of points. In this pattern, it is possible to observe lighter lines, called interference fringes, which are the result of the constructive interference. Figure 2 shows two wave fronts and the resulting interference pattern, where the interference fringes are clearly observed.

3.2. *Phase maps*

In shearography, the phase maps, which are obtained from the speckle patterns, contain information about the displacement derivatives of a surface. Therefore, if we are able to quantify this information, we can directly obtain the values of the displacement derivatives. The speckle pattern is formed due to the illumination of the rough surface by laser light and can be recorded in a digital camera with CCD or CMOS sensors.

The fundamental equations of shearography, which relate the phase maps with displacement derivatives of a surface, can be derived from vector theory.[26] In a Cartesian set of coordinates, the displacement fields of a surface can be represented by the vectors $\vec{u}(x,y)$, $\vec{v}(x,y)$ and $\vec{w}(x,y)$, describing the displacements in the x, y, and z directions, respectively. Let us consider a shearing amount Δx in the x direction, which is defined by the distance between any two points $P_1(x,y)$ and $P_2(x + \Delta x, y)$, and a sensitivity vector $\vec{k}$ with an average angle $\theta_{xz}/2$ between the illumination and observation vectors (Figure 3). It can be shown that the phase map

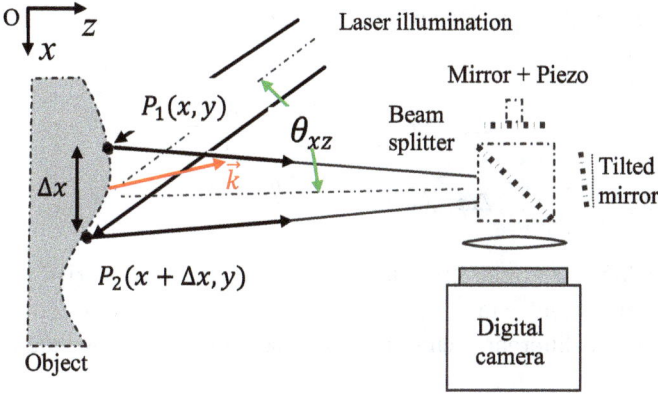

Figure 3. Geometric relations in shearography with a Michelson interferometer.

$\Delta\phi_{xx}(x, y)$, that are defined in the range $[-\pi, \pi]$, is given by[26]

$$\Delta\phi_{xx}(x, y) = \frac{2\pi\Delta x}{\lambda} \left\{ \sin(\theta_{xz}) \frac{\partial u(x, y)}{\partial x} + [1 + \cos(\theta_{xz})] \frac{\partial w(x, y)}{\partial x} \right\} \quad (16)$$

If the shear is considered in the y direction, a different phase map is obtained

$$\Delta\phi_{yy}(x, y) = \frac{2\pi\Delta y}{\lambda} \left\{ \sin(\theta_{xz}) \frac{\partial u(x, y)}{\partial y} + [1 + \cos(\theta_{xz})] \frac{\partial w(x, y)}{\partial y} \right\} \quad (17)$$

Analogous equations are obtained if the illumination plane is yz, and thus the sensitivity vector is characterized by an angle $\theta_{yz}/2$, being the system sensitive to motion in the plane yz

$$\Delta\phi_{xx}(x, y) = \frac{2\pi\Delta x}{\lambda} \left\{ \sin(\theta_{yz}) \frac{\partial v(x, y)}{\partial x} + [1 + \cos(\theta_{yz})] \frac{\partial w(x, y)}{\partial x} \right\} \quad (18)$$

$$\Delta\phi_{yy}(x, y) = \frac{2\pi\Delta y}{\lambda} \left\{ \sin(\theta_{yz}) \frac{\partial v(x, y)}{\partial y} + [1 + \cos(\theta_{yz})] \frac{\partial w(x, y)}{\partial y} \right\} \quad (19)$$

The four equations above show that phase map contains information about the in-plane components u (or v) and the out-of-plane component w. When the illumination is such that $\theta_{xz} = 0$ or $\theta_{yz} = 0$, the in-plane components vanish and Equations (16) and (17), or (18) and (19), respectively, become

$$\Delta\phi_{xx}(x,y) = \frac{4\pi\Delta x}{\lambda}\frac{\partial w(x,y)}{\partial x} \tag{20}$$

$$\Delta\phi_{yy}(x,y) = \frac{4\pi\Delta y}{\lambda}\frac{\partial w(x,y)}{\partial y} \tag{21}$$

Equations (20) and (21) clearly show that the derivative of the out-of-plane component can be measured from the phase maps, which can be determined by different techniques as presented in the following sections.

3.2.1. *Temporal phase shifting*

The interferometer used to obtain the speckle pattern relies on two interfering wave fronts, which are laterally sheared and come directly from the surface of the object under investigation. The sheared wave fronts can be created using a Michelson interferometer with a slightly tilted mirror. This type of interferometer allows an easy adjustment of the wave front lateral shear and the application of the phase shifting or phase stepping technique to extract the phase maps. In the first step of this technique, we obtain the interference phases of the reference and deformed states of the surface, respectively, $\Phi_R(x,y)$ and $\Phi_D(x,y)$, based on at least three speckle patterns for each state. Usually, the reference state corresponds to the case where no excitation is applied to the object, whereas the deformed state in the context of modal analysis corresponds to the maximum amplitude of vibration. The need for at least three patterns is related to the number of unknowns that must be extracted from the intensities. One of the most used techniques is based on four recorded patterns, where a phase shift of $\pi/2$ is defined between each recorded pattern, such that $\Phi_R(x,y)$ and $\Phi_D(x,y)$ are given, respectively, by

$$\Phi_R(x,y) = \arctan\left[\frac{I_{R,4}(x,y) - I_{R,2}(x,y)}{I_{R,1}(x,y) - I_{R,3}(x,y)}\right] \tag{22}$$

and

$$\Phi_D(x,y) = \arctan\left[\frac{I_{D,4}(x,y) - I_{D,2}(x,y)}{I_{D,1}(x,y) - I_{D,3}(x,y)}\right] \tag{23}$$

where the first subscript denotes the different states, whereas the second indicates the phase shift order of the different recorded intensities.

The second and last step leads to the computation of the phase map $\Delta\phi(x, y)$, which is just the subtraction of the previously obtained interference phases in the two states

$$\Delta\phi(x, y) = \begin{cases} \Phi_D(x, y) - \Phi_R(x, y) - \pi & \text{if } \Phi_D \geq \Phi_R(x, y) \\ \Phi_D(x, y) - \Phi_R(x, y) + \pi & \text{if } \Phi_D < \Phi_R(x, y) \end{cases} \tag{24}$$

Alternative temporal phase shifting methods are extensively described in Ref. 61. The selection of the method is based on the function of its efficiency in dealing with miscalibrations of the optical setup and the computational effort required. Figure 4 shows schematically how the phase map of a fully clamped plate subjected to a uniform pressure is obtained with the method of temporal phase shifting with four shifts or steps of $\pi/2$, as described by Equations (22)–(24).

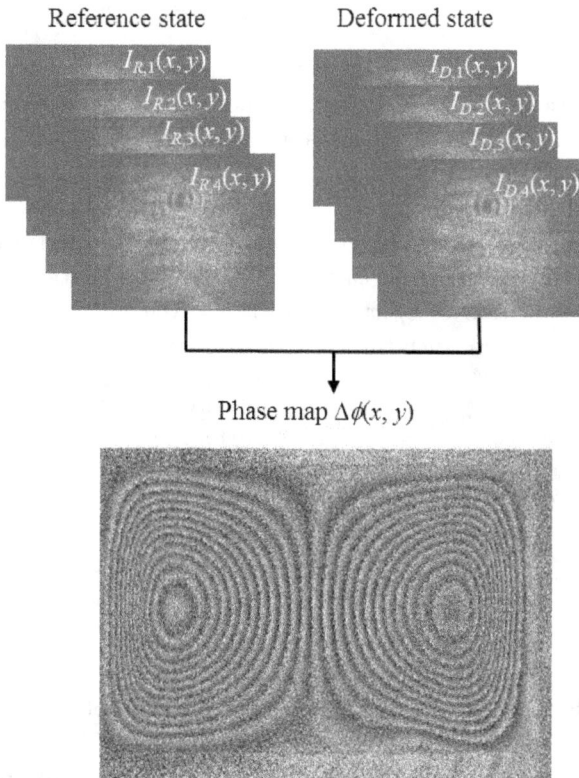

Figure 4.   Illustration of the process of phase map extraction from speckle pattern intensities with temporal phase shifting.

3.2.2. *Spatial carrier*

The phase map can also be obtained when only a recorded intensity distribution is available in each state. In this case, the approach is based on the introduction of a spatial carrier in the intensity distribution. The spatial carrier can be created by a slight tilt of one of the mirrors of a Mach–Zehnder optical configuration. The interference phases of the reference $\Phi_R(x, y)$ and deformed states $\Phi_D(x, y)$ are obtained by isolating the spectral information around the spatial carrier. This is accomplished by applying sequentially a fast Fourier transform (FFT) and an inverse fast Fourier transform (IFFT).[65] Let us consider the intensity of the interference in the reference state $\hat{I}_R(u, v)$. This intensity in the wave number domain can be defined as a function of the background intensity $\hat{A}_R(u, v)$ and the carrier intensity $\hat{C}_R(u, v)$, which is modeled by the interference intensity:

$$\hat{I}_R(u, v) = \hat{A}_R(u, v) + \hat{C}_R(u, v) + \hat{C}_R^*(u, v) \tag{25}$$

where u and v are the order of the wave number in the horizontal and vertical directions, respectively, and $*$ denotes the complex conjugate. The intensity of the phase interference is separated from the background intensity by the application of a window filter. The spatial carrier is adjusted by controlling the rotation of one of the mirrors in the optical setup, whereas the window filter is adjusted by controlling its optical aperture. The interference phase $\Phi_R(x, y)$ can be calculated, after demodulation of the spatial carrier, as a function of $c_R(x, y)$, which is given by the inverse Fourier transform of $\hat{C}_R(u, v)$, such that

$$\Phi_R(x, y) = \arctan \frac{\mathrm{Im}[c_R(x, y)]}{\mathrm{Re}[c_R(x, y)]} \tag{26}$$

The intensity of the interference in the deformed state $\hat{I}_D(u, v)$ is obtained in a similar fashion, such that the interference phase of the deformed state is given by

$$\Phi_D(x, y) = \arctan \frac{\mathrm{Im}[c_D(x, y)]}{\mathrm{Re}[c_D(x, y)]} \tag{27}$$

Figure 5 presents the different steps needed to obtain the final phase map using the spatial modulation technique described. The example reports the study of a mode shape of a composite plate completely free. The last step is the same as the last one in the temporal phase shifting technique, and thus the phase map is computed by Equation (24).

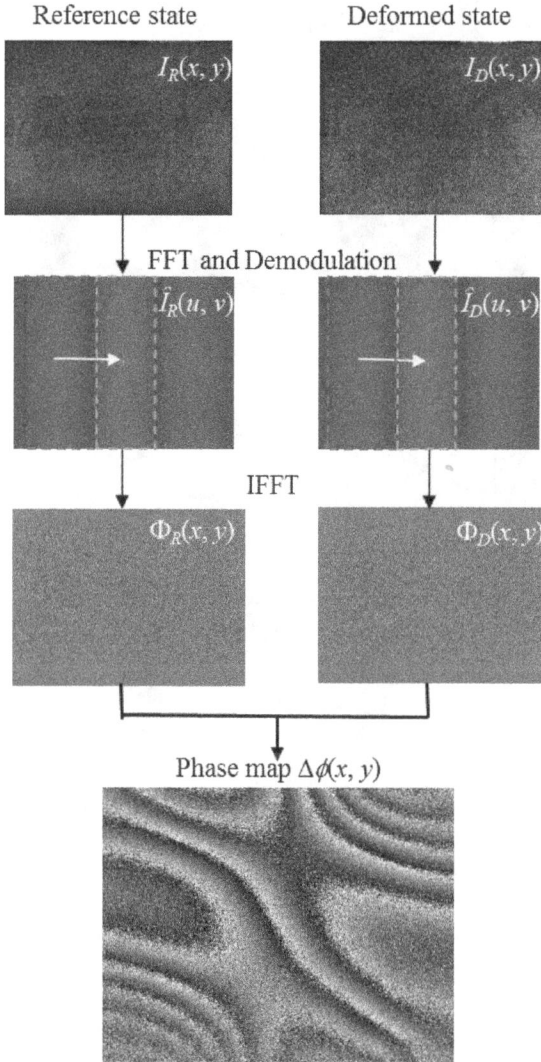

Figure 5.   Illustration of the process of phase map extraction from speckle pattern intensities with spatial carrier modulation.

3.3. *Post-processing of phase maps*

3.3.1. *Filtering*

As can be seen in Figures 4 and 5, the phase maps present large levels of high frequency noise. A commonly used filter is the average low-pass filter,

(a)   (b)

(c)

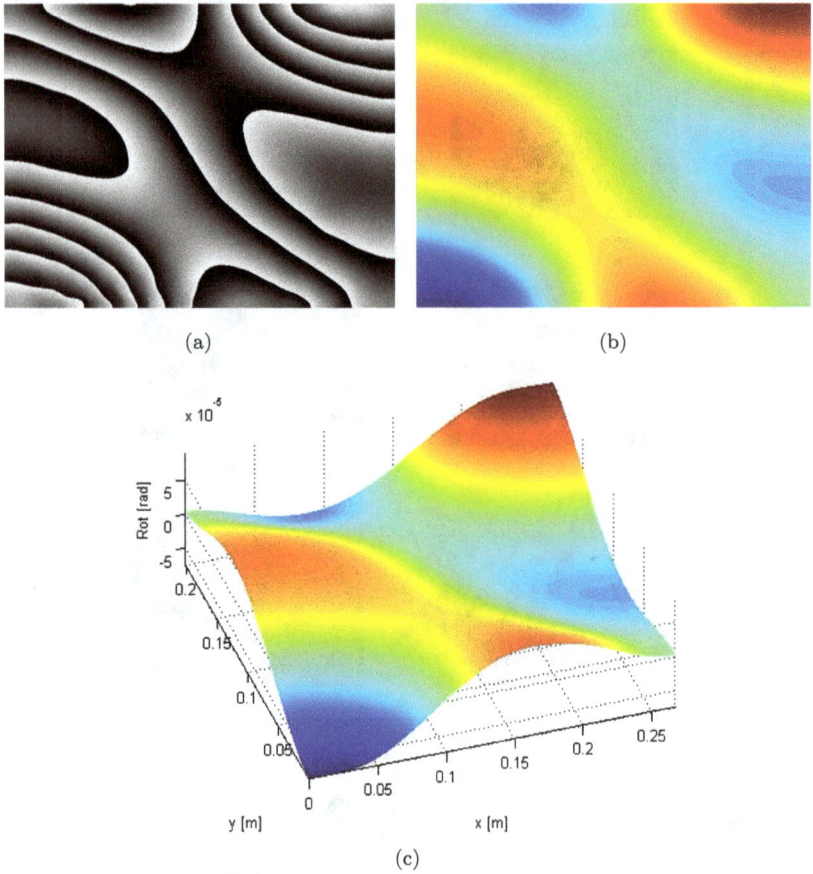

Figure 6.   Illustration of the process of phase map post-processing: (a) filtering, (b) unwrapping, and (c) 3D representation of mode shape gradient or modal rotation field.

although other more complex and sophisticated filters can be applied to eliminate the noise more effectively. A comparison of nine different filtering techniques was carried out and described in Refs. 66 and 67. The filters based on the windowed Fourier transform and short-time Fourier transform, which are applied in the wave number domain, show the best performance, although at the expense of great computational effort. For instance, the computational time for the processing of a phase map with the windowed Fourier transform is almost 15 folds the time necessary for the same task with the average filter.[66] The filtered phase map described in the last image of Figure 5 is presented in Figure 6(a).

3.3.2. *Unwrapping*

Due to the very nature of the process of their computation, the phase maps are discontinuous. Indeed, they are only defined in the interval $[-\pi, \pi]$ and we say that the phase maps are wrapped. In order to remove these discontinuities, and thus to obtain the continuous fields of interest, we must apply unwrapping methods. The success of these methods relies on the correct identification of the phase discontinuities, this being accomplished by previously applying filtering techniques to the phase maps. The basic idea behind unwrapping methods is the addition or subtraction of 2π to values of phase below or above $-\pi$ and π, respectively. However, because the phase map presents ambiguities and/or inconsistencies, this simple process may fail and more complex unwrapping methods must be applied. Two main types of methods are available, known as path-following and minimum norm methods.[68] In the first case, we determine the continuous phase by integrating along a path defined by restriction lines or branch cuts. The minimum norm methods are based on the minimization of an error norm given by the difference between the continuous field and the measured discontinuous phase map. Refs. 66 and 67 present a study on the performance of eight different unwrapping methods. This study shows that the path-following methods are more effective in dealing with phase discontinuities, whereas the minimum norm methods are more efficient in the treatment of inconsistencies. Reference 68 extensively explains the importance of phase unwrapping in different types of interferometric techniques, presenting also several algorithms and their computational implementation.

Figure 6 shows the process of post-processing of the phase map presented in Figure 5. In Figure 6(a), we can see the filtered phase map, in Figure 6(b), the unwrapped phase map, coded in colors, and in (c), a 3D representation of the mode shape gradient or modal rotation field.

4. Damage Localization

A clamped–clamped aluminum beam with a length of 351 mm, a width of 40 mm and a thickness of 2.1 mm was subjected to three cases of damages. Each case of damage is characterized by a saw cut through the width at coordinate $x = (x_1 + x_2)/2 = 90$ mm (Figure 7). All the cuts have a length of $c = 1$ mm, but have varying depths of $p = 0.3$, 0.5 and 1.0 mm.

The beam was excited acoustically at its three first natural frequencies and the corresponding three modal rotation fields are obtained

J. V. Araújo dos Santos & H. Lopes

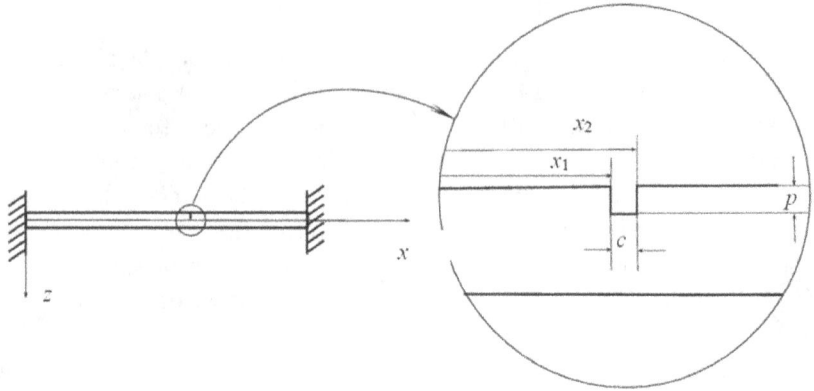

Figure 7.   Dimensions and localization of damage on a clamped–clamped beam.

with shearography, based on a Michelson interferometer with strobo-scopic illumination. This stroboscopic illumination is achieved by an acousto-optic modulator that chops the light of a continuous wave laser with the exciting frequency.[26] The phase maps are evaluated with tempo-ral phase shifting and are filtered and unwrapped in order to obtain the continuous fields (Figure 8). From these continuous fields, we can obtain 1D profiles of the modal rotations $\theta(x)$, which are taken from the mid cross-section of the beam in the width direction. These modal rotation profiles are afterwards used for damage evaluation. The damage indica-tors used are the summation over the first three modes of the modi-fied curvature difference absolute values (MCD) and the modified damage index (MDI)

$$\text{MCD}(i, x_l) = \left| \frac{d\tilde{\theta}_i(x_l)}{dx} - \frac{d\theta_i(x_l)}{dx} \right| \tag{28}$$

$$\text{MDI}(i, x_l) = \frac{\left[\left(\frac{d\tilde{\theta}_i(x_l)}{dx} \right)^2 + \sum_{k=1}^{NP} \left(\frac{d\tilde{\theta}_i(x_k)}{dx} \right)^2 \right] \sum_{k=1}^{NP} \left(\frac{d\theta_i(x_k)}{dx} \right)^2}{\left[\left(\frac{d\theta_i(x_l)}{dx} \right)^2 + \sum_{k=1}^{NP} \left(\frac{d\theta_i(x_k)}{dx} \right)^2 \right] \sum_{k=1}^{NP} \left(\frac{d\tilde{\theta}_i(x_k)}{dx} \right)^2} \tag{29}$$

where i denotes the mode shape, x_l is the coordinate where the spatial derivative of the maximum amplitudes of the undamaged and damaged rotations $\theta(x)$ and $\tilde{\theta}(x)$, respectively, and NP is the number of measured points. In this case, the number of measured points is equal to 2158, which is the number of pixels in the x direction. The undamaged rotations

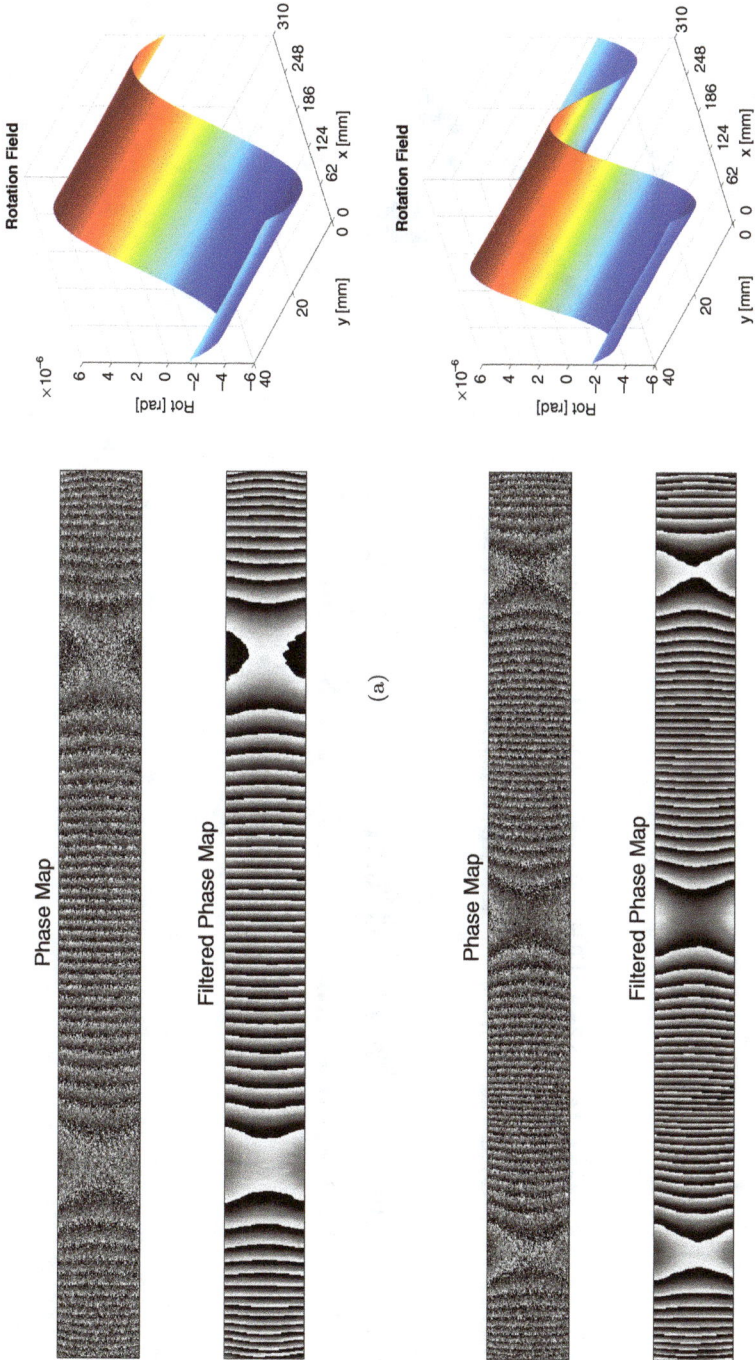

Figure 8. Phase maps, filtered phase maps and 3D representations of modal rotation fields of, (a) first, (b) second, and (c) third modes of aluminum beam.

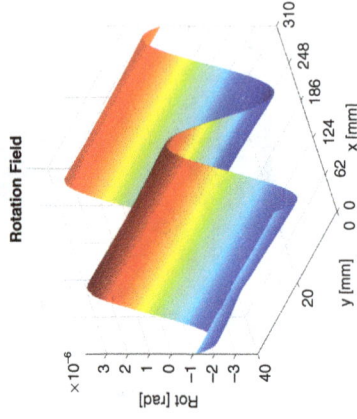

(c)   (*Continued*)

Figure 8.

were computed by the Ritz method and the Timoshenko theory.[69,70] Equations (28) and (29) are modifications of the curvature difference proposed by Pandey *et al.*[1] and the DI proposed by Stubbs *et al.*,[2] respectively, because in the present case, we differentiate only once the rotations and not twice the displacements.

While the undamaged beam rotations can be easily computed by analytical differentiation of the assumed functions in the Ritz method, the damage beam rotations are differentiated by image convolution between experimental data and the 1D first-order Gaussian derivative:

$$\frac{\partial \tilde{\theta}(x)}{\partial x} = \frac{\tilde{\theta}(x) \otimes \bar{\partial}\bar{G}(i)}{\Delta x} = \frac{F^{-1}[F(\tilde{\theta}(x)) \times F(\bar{\partial}\bar{G}(i))]}{\Delta x} \qquad (30)$$

where $\otimes$ is the convolution symbol, F and F^{-1} represent the direct and inverse Fast Fourier transform, respectively, Δx the distance between two consecutive points and the Gaussian first-order derivative kernel $\bar{\partial}\bar{G}(i)$ is assumed to be extended by zero padding. The normalized Gaussian first-order derivative kernel, with a width of nine points, can be computed considering the following equation[71]:

$$\bar{d}\bar{G}(i) = \frac{dG(i)}{\sum_{j=-4}^{4} j\,dG(j)} = \frac{-i}{\sqrt{2\pi}} e^{-i^2/2} \quad \text{with } i = -4, -3, \ldots, 3, 4 \quad (31)$$

Figure 9 presents the damage indicators along the coordinate x. Near the extremities of the beam, the damage indicators present maximum or minimum values for all cases of damage. Therefore, this is a systematic behavior and cannot be justified by the presence of damage. This is mainly due to difficulties in accomplishing a perfect clamping of the beam. We see that the cut with the lowest depth ($p = 0.3\,\text{mm}$) does not present a noticeably high value of the damage indicators in coordinate $x = 0.09\,\text{m}$, but it is clear that a peak in the curves is observed for the cuts with depths $p = 0.5$ and $1.0\,\text{mm}$.

As presented in Ref. 50, multiple damage in an aluminum beam can also be localized with relative success using a pulsed speckle shearography system, which is based on a Mach–Zehnder interferometer, where the phase maps are obtained with spatial phase modulation. The damages are saw cuts with depths higher than 1/10 of the thickness. More recently, a new optical setup of a Michelson interferometer, which relies on stroboscopic illumination and temporal phase shifting, allowed obtaining even more accurate measurements.[53] The system was used to test its efficiency in the localization of multiple damage in a free–free aluminum beam with

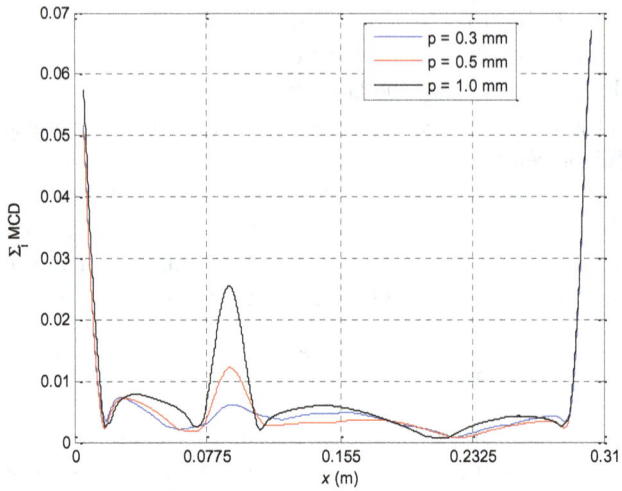

(a)

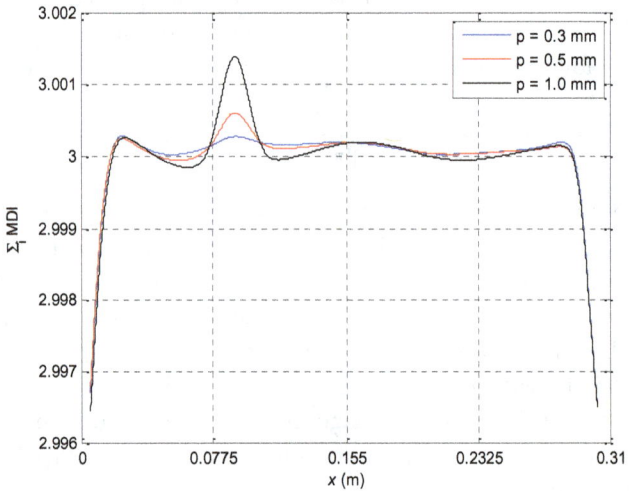

(b)

Figure 9.   Summation over the first three modes of (a) the MCD and (b) the MDI.

dimensions $400 \, \text{mm} \times 40 \, \text{mm} \times 3 \, \text{mm}$. The damages were created by a milling machine, such that slots are located at mid-span and at $284 \, \text{mm}$ from the left edge, as listed in Table 1. In order to verify the result repeatability, two damage scenarios were studied. The first damage scenario has a very small slot at mid-span, with a depth equal to $1/100$ of the thickness. The second

Table 1. Damage scenarios with the positions and dimensions of the slots.

Damage scenario	Slot	Position (mm)	Width (mm)	Depth (mm)
1	1	200	3	0.03
	2	284	5	0.41
2	1	200	3	0.30
	2	284	5	0.41

damage scenario is such that both slots have a depth of at least 1/10 of the thickness, thus describing a relatively small damage.

The analytical and experimental modal rotation field profiles of the third mode and their first, second and third-order derivatives are presented in Figure 10. The experimental derivatives were computed by applying in succession the image convolution technique described in Equation (30) to the measured data obtained with shearography. The analytical profiles are the ones obtained from the analytical solution of a free vibrating uniform beam, i.e. of a beam without slots and with a constant thickness.

It is clear from Figure 10(a) that the modal rotation field profiles of the two damage scenarios are very smooth and similar to the analytical profile, and thus it is not possible to localize the slots. As can be seen in Figure 10(b), with the first-order derivative of the modal rotation field, which corresponds to the curvature of the mode shape, we start to observe perturbations in the curves for the slots with the largest depths. This proves that with the first-order derivative of the modal rotation field profiles, one is only able to localize the largest slots in both damage scenarios and the slot at mid-span when it has a depth of 1/10 of the thickness. It is worth mentioning that the application of central finite differences to obtain the first derivative of the modal rotation field profiles, which is reported in Ref. 53, also only leads to the localization of the largest slot, but not the smallest for the first damage scenario in the present work.

For the second-order derivative of the modal rotation field, the perturbations at the locations of the slots are larger and clearer, although once again the smallest slot at mid-span is not localized, as can be seen in Figure 10(c). With the third-order derivative of the modal rotation field, a small perturbation is observed at mid-span, indicating the presence of the smallest slot of the first damage scenario, as illustrated in Figure 10(d). It should be stressed that this slot has a depth that is 100 times smaller than the thickness of the beam. The slot in the mid-span of damage scenario two has a depth one order of magnitude larger and, for this case, the peak of the perturbations is about eight times higher than the peak of the perturbations

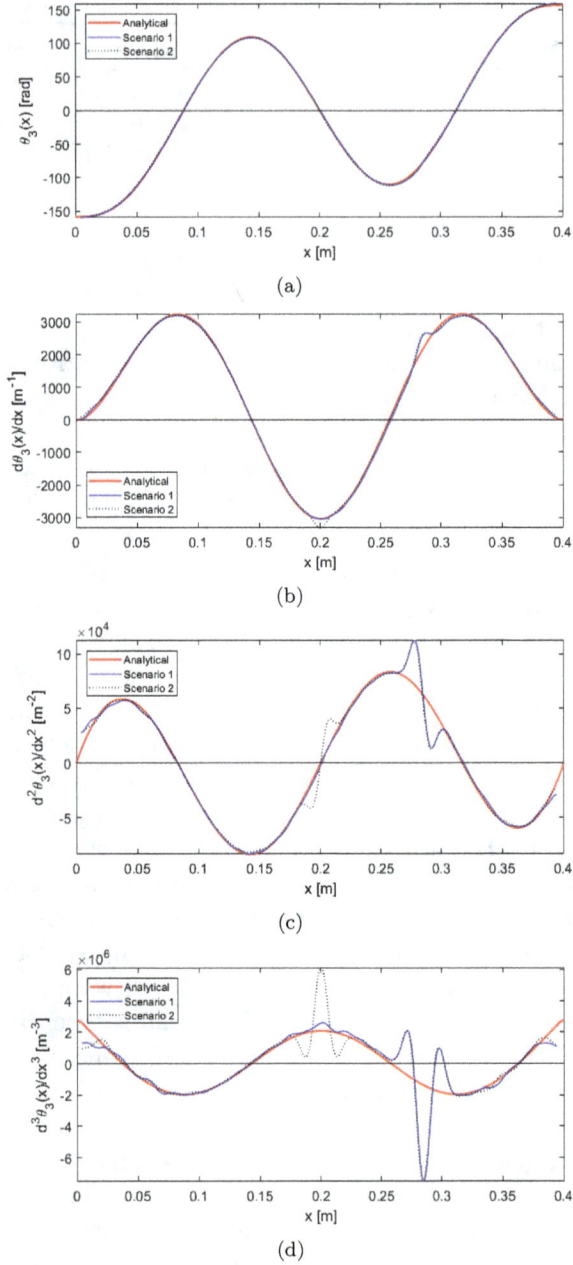

Figure 10. Analytical and experimental (a) modal rotation field profiles of the third mode and their (b) first-, (c) second-, and (d) the third-order derivatives of a free–free aluminum beam.

of the slot in mid-span of damage scenario one. As an overall conclusion, the present results thus show that the use of higher order derivatives may be an interesting alternative in order to detect small damages.

Another noticeable feature of the present results is that an excellent repeatability is observed. Indeed, in addition to the similarity in the global trend of the curves in Figure 10, the perturbations due to the second slot in both damage scenarios are clearly similar, proving that although the measurements were taken at different times, they do not differ much. Therefore, shearography can be considered as a very stable measurement technique.

5. Conclusions

Shearography, which is a non-contact, full-field, and high-resolution optical method, is described in this chapter. It is also shown that this method can be used as an alternative to usual modal analysis techniques. In addition to a review of the literature containing applications of shearography for the measurement of vibrating objects, several studies of damage localization are also reported. The structures analyzed are aluminum beams, and it is found that very small damage can be localized, namely with higher order derivatives of the measured modal rotation fields. Furthermore, multiple damages can also be localized accurately and effectively, thus showing that shearography is a very powerful technique for the experimental measurement of modal response of structures. The applications reported in this chapter also illustrate that new developments of numerical–experimental methods are opened to be explored by applying full-field techniques, such as shearography, as well as improvements in usual methods for detection, localization, and quantification of structural damage.

Acknowledgements

This work was supported by FCT, through IDMEC, under LAETA, Project UID/EMS/50022/2013.

References

1. Pandey, A., Biswas, M., and Samman, M. Damage detection from changes in curvature mode shapes. *Journal of Sound and Vibration* **145**, pp. 321–332 (1991).
2. Stubbs, N., Kim, J. T., and Farrar, C. R. Field verification of a nondestructive damage localization and severity estimator algorithm. In *Proceedings of the 13th IMAC*, pp. 210–218 (1995).

3. Ratcliffe, C. Damage detection using a modified laplacian operator on mode shape data. *Journal of Sound and Vibration* **204**, pp. 505–517 (1997).
4. Sampaio, R. P. C., Maia, N. M. M., and Silva, J. M. M. Damage detection using the frequency-response-function curvature method. *Journal of Sound and Vibration* **226**, pp. 1029–1042 (1996).
5. Sazonov, E. and Klinkhachorn, P. Optimal spatial sampling interval for damage detection by curvature or strain energy mode shapes. *Journal of Sound and Vibration* **285**, pp. 783–801 (2005).
6. Guan, H. and Karbhari, V. Improved damage detection method based on element modal strain damage index using sparse measurement. *Journal of Sound and Vibration* **309**, pp. 465–494 (2008).
7. Chandrashekhar, M. and Ganguli, R. Damage assessment of structures with uncertainty by using mode-shape curvatures and fuzzy logic. *Journal of Sound and Vibration* **326**, pp. 939–957 (2009).
8. Rucevskis S. and Wesolowski, M. Identification of damage in a beam structure by using mode shape curvature squares. *Shock Vibration* **17**, pp. 601–610 (2010).
9. Tomaszewska, A. Influence of statistical errors on damage detection based on structural flexibility and mode shape curvature. *Computers and Structures* **88**, pp. 154–164 (2010).
10. Radzieński, M., Krawczuk, M., and Palacz, M. Improvement of damage detection methods based on experimental modal parameters. *Mechanical Systems and Signal Processing* **25**, pp. 2169–2190 (2011).
11. Solis, M., Algaba, M., and Galvan, P. Continuous wavelet analysis of mode shapes differences for damage detection. *Mechanical Systems and Signal Processing* **40**, pp. 645–666 (2013).
12. Cao, M., Xu, W., Ostachowicz, W., and Su, Z. Damage identification for beams in noisy conditions based on Teager energy operator-wavelet transform modal curvature. *Journal Sound Vibration* **333**, pp. 1543–1553 (2014).
13. Xu, Y., Zhu, W., Liu J., and Shao, Y. Identification of embedded horizontal cracks in beams using measured mode shapes. *Journal of Sound and Vibration* **333**, pp. 6273–6294 (2014).
14. Yang, C., Fu, Y., Yuan, J., Guo, M., Yan, K., Liu, H., Miao, H., and Zhu, C. Damage identification by using a self-synchronizing multipoint laser Doppler vibrometer. *Shock Vibration* **2015**, p. 9 (2015).
15. Sirohi, R. S. *Optical Methods of Measurement: Wholefield Techniques*, 2nd edn. CRC Press, Boca Raton (2009).
16. Patsias, S. and Staszewski, W. J. Damage detection using optical measurements and wavelets. *Structural Health Monitoring* **1**, pp. 5–22 (2002).
17. Araújo dos Santos, J. V., Lopes, H. M. R., Vaz, M., Soares, C. M. M., Soares, C. A. M., and de Freitas, M. J. M. Damage localization in laminated composite plates using mode shapes measured by pulsed TV holography. *Composite Structures* **76**, pp. 272–281 (2006).
18. Helfrick, M., Pingle, P., Niezrecki, C., and Avitabile, P. Using full-field vibration measurement techniques for damage detection. In *Proceeding of the 27th IMAC* (2009).

19. Dworakowski, Z., Kohut, P., Gallina, A., Holak, K., and Uhl, T. Vision-based algorithms for damage detection and localization in structural health monitoring. *Structural Control and Health Monitoring* **23**, pp. 35–50 (2016).

20. Leendertz J. A. and Butters, J. N. An image-shearing speckle-pattern interferometer for measuring bending moments. *Journal Physics E: Scientific Instruments* **6**, pp. 1107–1110 (1973).

21. Vlasov, N. G. and Presnyakov, Y. P. Shearing interferometry of diffusely reflecting objects. *Soviet Journal Quantum Electron* **3**, pp. 141–143 (1973).

22. Hung, Y. Y. and Taylor, C. E. Speckle-shearing interferometric camera — a tool for measurement of derivatives of surface displacements. In *Proceeding SPIE 41*, pp. 169–175 (1974).

23. Butters, J. N. and Leendertz, J. A. Speckle pattern and holographic techniques in engineering metrology. *Optics and Laser Technology* **3**, pp. 26–30 (1971).

24. Hariharan, P. Speckle patterns: A historical retrospect. *Optics Acta* **19**, pp. 791–793 (1972).

25. Dainty, J. C. (ed.). *Laser Speckle and Related Phenomena.* Springer-Verlag, Berlin Heidelberg GmbH (1975).

26. Steinchen, W. and Yang, L. *Digital Shearography: Theory and Application of Digital Speckle Pattern Shearing Interferometry*, SPIE Press, Bellingham, Washington (2003).

27. Cloud, G. Practical speckle interferometry for measuring in-plane deformation. *Applied Optics* **14**, pp. 878–884 (1975).

28. Archbold, E., Burch, J. M., Ennos, A. E., and Taylor, P. A. Visual observation of surface vibration nodal patterns. *Nature* **222**, pp. 263–265 (1969).

29. Ek, L. and Molin, N.-E. Detection of the nodal lines and the amplitude of vibration by speckle interferometry. *Optics Communications* **2**, pp. 419–424 (1971).

30. Tiziani, H. J. Analysis of mechanical oscillations by speckling. *Applied Optics* **11**, pp. 2911–2917 (1972).

31. Fernelius N. and Tome, C. Vibration-analysis studies using changes of laser speckle. *Journal of the Optical Society of America* **61**, pp. 566–572 (1971).

32. Hung, Y. Y. Digital shearography versus TV-holography for non-destructive evaluation. *Optical Laser Engineering* **26**, pp. 421–436 (1997).

33. Hung, Y. Y., Rowlands, R. E., and Daniel, I. M. Speckle-shearing interferometric technique: A full-field strain gauge. *Applied Optics* **14**, pp. 618–622 (1975).

34. Chiang, F. P. and Juang, R. M. Vibration analysis of plate and shell by laser speckle interferometry. *Optics Acta* **23**, pp. 997–1009 (1976).

35. Hung, Y. Y., Displacement and strain measurement, In ed. R. K. Erf, *Speckle Metrology.* Academic Press, New York (1978).

36. Mohan, N. K., Saldner H., and Molin, N.-E. Electronic shearography applied to static and vibrating objects. *Optics Communications* **108**, pp. 197–202 (1994).

37. Toh, S., Tay, C., Shang, H., and Lin, Q. Time-average shearography in vibration analysis. *Optics and Laser Technology* **27**, pp. 51–55 (1995).

38. Sim, C., Chau, F., and Toh, S. Vibration analysis and non-destructive testing with real-time shearography. *Optics and Laser Technology* **27**, 45–49 (1995).
39. Hung, Y. Y. and Ho, H. P. Shearography: An optical measurement technique and applications. *Materials Science and Engineering R* **49**, pp. 61–87 (2005).
40. Nakadate, S., Yatagai, T., and Saito, H. Digital speckle-pattern shearing interferometry. *Applied Optics* **19**, pp. 4241–4246 (1980).
41. Ng, T. W. and Chau, F. S. A digital shearing speckle interferometry technique for modal analysis. *Applied Acoustics* **42**, pp. 175–185 (1994).
42. Yang, L., Steinchen, W., Kupfer, G. Mäckel, P., and Vössing, F. Vibration analysis by means of digital shearography. *Optics and Laser Engineering* **30**, pp. 199–212 (1998).
43. Wong W. O. and Chan, K. T. Measurement of modal damping by electronic speckle shearing interferometry. *Optics and Laser Technology* **30**, pp. 113–120 (1998).
44. Casillas, F. J., Dávila, A., Rothberg, S. J., and Garnica, G. Small amplitude estimation of mechanical vibrations using electronic speckle shearing pattern interferometry. *Optical Engineering* **43**, pp. 880–887 (2004).
45. Bhaduri, B., Kothiyal, M. P., and Mohan, N. K. Vibration mode shape visualization with dual function DSPI system. In *Proceeding SPIE 6292*, Vol. 2006, pp. 629217-1–629217-7 (2006).
46. Valera, J. D. R. and Jones, J. D. C. Vibration analysis by modulated time-averaged speckle shearing interferometry. *Measuremental Science and Technology* **6**, pp. 965–970 (1995).
47. Somers, P. A. A. M. and Bhattacharya, N. Vibration phase-based ordering of vibration patterns acquired with a shearing speckle interferometer and pulsed illumination. *Strain* **46**, pp. 234–241 (2010).
48. Steinchen, W., Yang, L., Kupfer, G., and Mäckel, P. Non-destructive testing of aerospace composite materials using digital shearography. In *Proceedings of the Institutional of Mechanical Engineering G-J. Aerospace Engineering*, Vol. 212, pp. 21–30 (1998).
49. Hung, Y. Y., Luo, W. D., Lin, L., and Shang, H. M. Evaluating the soundness of bonding using shearography. *Composite Structures* **50**, pp. 353–362 (2000).
50. Lopes, H. M. R., Ribeiro, J., and Araújo dos Santos, J. V. Interferometric techniques in structural damage identification. *Shock Vibration* **19**, pp. 835–844 (2012).
51. Lopes, H. M. R., Araújo dos Santos, J. V., Soares, C. M. M., Guedes, R. J. M., and Vaz, M. A. P. A numerical-experimental method for damage location based on rotation fields spatial differentiation. *Composite Structures* **89**, pp. 1754–1770 (2011).
52. de Medeiros, R., Lopes, H. M. R., Guedes, R. J. M., Vaz, M. A. P., Vandepitte, D., and Tita, V. A new methodology for structural health monitoring applications. *Procedia Engineering* **114**, pp. 54–61 (2015).
53. Mininni, M., Gabriele, S., Lopes, H. M. R., and Araújo dos Santos, J. V. A. Damage identification in beams using speckle shearography and an optimal spatial sampling. *Mechanical Systems and Signal Processing* **79**, pp. 47–64 (2016).

54. Chen, F. Digital shearography: State of the art and some applications. *Journal of Electronic Imaging* **10**, pp. 240–251 (2001).

55. Francis, D., Tatam, R. P., and Groves, R. M. Shearography technology and applications: A review. *Measurements Science and Technology* **21**, 102001 (2010).

56. Hung, Y. Y., Chen, Y. S., Ng, S. P., Liu, L., Huang, Y. H., Luk, B. L., Ip, R. W. L., Wu, C. M. L., and Chung, P. S. Review and comparison of shearography and active thermography for nondestructive evaluation. *Materials Science and Engineering Reports* **64**, pp. 73–112 (2009).

57. Garnier, C., Pastor, M.-L., Eyma, F., and Lorrain, B. The detection of aeronautical defects in situ on composite structures using Non Destructive Testing. *Composite Structures* **93**, pp. 1328–1336 (2011).

58. Amenabar, I., Mendikute, A., López-Arraiza, A., Lizaranzu M., and Aurrekoetxea, J. Comparison and analysis of non-destructive testing techniques suitable for delamination inspection in wind turbine blades. *Composites Part B: Engineering* **42**, pp. 1298–1305 (2011).

59. Zastavnik, F., Pyl, L., Gu, J., Sol, H., Kersemans, M., and Van Paepegem, W. Comparison of shearography to scanning laser vibrometry as methods for local stiffness identification of beams. *Strain* **50**, pp. 82–94 (2014).

60. Froehly, C. Speckle Phenomena and some of its applications, In ed. A. Lagarde, *Optical Methods in Mechanics of Solids*. Sijthoff & Noordhoff (1980).

61. T. Kreis, *Handbook of Holographic Interferometry: Optical and Digital Methods*. Wiley-VCH (2005).

62. Gan, Y. and Steinchen, W. Chaper 23: Speckle Methods, In ed. W. N. Sharpe, *Springer Handbook of Experimental Solid Mechanics*, Springer (2008).

63. Mohan, N. K. Chapter 8: Speckle methods and applications, In ed. T. Yoshizawa. *Handbook of Optical Metrology: Principles and Applications*. CRC Press (2008).

64. Gåsvik, K. J. *Optical Metrology*. John Wiley & Sons, Chichester, West Sussex (2002).

65. Takeda, M., Ina, H., and Kobayashi, S. Fourier-transform method of fringe-pattern analysis for computer-based topography and interferometry. *Journal of Optical Society of America* **72**, pp. 156–160 (1982).

66. Lopes, H. M. R. *Desenvolvimento de Técnicas Interferométricas, Contínuas e Pulsadas, Aplicadas à Análise do Dano em Estruturas Compósitas*, Ph.D. Thesis, Faculdade de Engenharia da Universidade do Porto, In Portuguese (2008).

67. Araújo dos Santos, J. V. and Lopes, H. Application of speckle interferometry to damage identification, In ed. B. Topping, *Computational Methods for Engineering Science*. Saxe-Coburg Publications, Stirlingshire, UK, pp. 299–330 (2012).

68. Ghiglia, D. C. and Pritt, M. D. *Two-Dimensional Phase Unwrapping: Theory, Algorithms, and Software*. Wiley, New York (1998).

69. Araújo dos Santos, J. V., Lopes, H. M. R., Ribeiro, J., Maia, N. M. M., and Vaz, M. A. P. Damage localization in beams using the Ritz method and

speckle shear interferometry, In eds. B. H. V. Topping, J. M. Adam, F. J. P. R. B., and Romero, M. L. In *Proceedings Tenth International Conference on Computational Structures Technology*. Civil-Comp Press (2010).

70. Araújo dos Santos, J. V., Lopes, H. M. R., and Maia, N. M. M. A damage localisation method based on higher order spatial derivatives of displacement and rotation fields. *Journal of Physics: Conference Series* **305**, p. 012008 (2011).

71. van den Boomgaard, R. and Smeulders, A. The morphological structure of images: The differential equations of morphological scale-space. *IEEE Transactions on Pattern Analysis and Machine* **16**, pp. 1101–1113 (1994).

Chapter 6

Novel Techniques for Damage Detection Based on Mode Shape Analysis

Wiesław Ostachowicz*,†,§, Maciej Radzieński*, Maosen Cao‡ and Wei Xu‡

*Institute of Fluid-Flow Machinery
Polish Academy of Sciences, 14 Fiszera St., 80-231 Gdansk, Poland
†Faculty of Automotive and Construction Machinery
Warsaw University of Technology, 02-524 Warsaw, Poland
‡Department of Engineering Mechanics, Hohai University
Nanjing 210098, People's Republic of China
§wieslaw.ostachowicz@imp.gda.pl

This chapter presents an overview of damage detection methods that make use of mode shapes. Both classic and novel approaches to mode shape analysis are presented. In the latter, two data sets (in the reference state and the tested state) are compared to extract information about potential changes caused by damage. In the former case, damage indices are based on various signal processing methods in which only one set of data is used for detecting and localizing any possible damage in a tested structure.

Keywords: Irregularity based damage detection; Mode shape analysis; Operational deflection shape (ODS); Modal assurance criterion (MAC); Coordinate modal assurance criterion (COMAC); Damage index (DI); Mode shape curvature (MSC); Strain energy method; Irregularity extraction; Modified Laplacian operator (MLO); Spectral modal curvature (SMC); Gapped smooth method; Unified load surface (ULS); Directional wavelets; Teager energy operator (TEO); Multiscale shear stain gradient; Fractal dimension.

1. Introduction

Mode shape-based damage detection techniques have been developed for a few decades now. Hundreds of papers can be found in the literature containing a number of different techniques and approaches for mode shape-based detection of damage. In this chapter, we present the most commonly used as well as the most promising methods, chosen on the basis of our experience in the field.

In the first part of this chapter, we present a brief overview of what is called the classic approach, where two sets of data (in the reference state and the tested state) are compared in order to reveal any possible changes that are related to structural health deterioration. In the second part, a more detailed overview is given of novel techniques that are based on various signal processing methods in which only one set of data is used for detecting and localizing any possible damage in a tested structure.

This chapter does not include methods that use any model updating technique for damage detection. It is important to note that the literature on the topic of mode shape-based damage detection is extensive, and so only chosen techniques are presented; this overview is by no means exhaustive.

It should be noted that for uniformity and to avoid any confusion in this chapter, the normalized form of an operation deflection shape (ODS) measured at or near the resonant frequency of a structure, which is dominated by one particular mode shape, is called a mode shape (MS). Full explanation of the differences between ODS and MS and the problem of what is actually measured in experimental studies can be found in Refs. 1 and 2, respectively.

2. Classic Mode Shape-Based Damage Detection Methods

West[3] was one of the first to reveal the possibility of using MSs for damage detection without modeling the tested structure. He used a modal assurance criterion (MAC) proposed by Allemang and Brown[4] to estimate correlations between experimental MSs of a space shuttle element in the reference state and after acoustic loading. Lieven and Ewins[5] extended the MAC, which gives information about the fact that the dynamics of a structure have been changed, to the coordinance modal assurance criterion (COMAC), which gives information about the part of a structure in which changes have occurred, and can be used as a damage index (DI).

Pandey *et al.*[6] in 1991 proposed using the mode shape curvature (MSC) as a tool for damage detection. They proved that changes in structural bending stiffness cause local changes in MSC that could be an appropriate tool for damage localization. The simplest and most commonly used method of calculating MSC is to use a numerical differentiation method: the central difference technique. Most papers have suggested using the second-order central difference algorithm, but a higher order algorithm can also be used, as presented by Qiao *et al.*[7]

Besides differences in the MSC of a structure between the pristine condition and the actual condition, differences in the mode shape slope or the mode shape curvature squared may be used, as compared in Refs. 8 and 9.

Cronwell[10] suggested dividing a structure into small fragments. A stiffness change in any of the small fragments can be used as a damage detection tool. This technique was called strain energy (SE) method. Stubbs *et al.*[11] proposed moving the ratio of the reference axis in order to avoid the possibility of its singularity occurrence. An approximate expression was used, where the reference point was moved from 0 to −1 for the healthy state of the tested structure. Their method was called the DI.

Choi *et al.*[12] proposed using an additional operation of normalization of every MSC to its maximum value. This method was called the modified damage index (MDI). This procedure provides higher effectiveness, especially in multiple damage scenarios.

From the abovementioned methods, those which are based on MSC (MSC, SE, DI, and MDI) are generally more effective in detecting small-sized damage. However, differentiation methods lead to an increase in the noise level. If no de-noising methods are used, the resulting increase in the number of points in measured mode shape can cause a decrease in the effectiveness of damage detection. At the same time, decreasing the number of points in MS leads to a truncation error and decreases the resolution of damage localization. This phenomenon was studied analytically by Sazanov and Klinkhachorn.[13] In that work, equations for calculating the optimal density of points for the SE and MSC methods are provided.

Another technique for damage localization involves the use of a flexibility matrix, which is the inverse of the stiffness matrix in order to estimate how the structure will behave under uniform static loading.[14] Based on this method and MSC, Zhang and Aktan[15] proposed to use flexibility curvature changes as a DI. More information about classic mode shape-based damage detection methods can be found in Ref. 16.

All the above-mentioned methods use two sets of measured MSs for inspection of a structure. The first set of data is obtained for the pristine condition (reference state) and the second set of data is measured in the actual/inspected condition state of the structure. In many cases, however, data for the pristine condition state is not available or measurement conditions cannot be properly repeated. This leads to a situation where only the current state of the structure can be used for health evaluation and possible damage detection and assessment. In contrast to the previously described "classic" methods, baseline-less methods can be used to locate any discontinuities in MSs caused by the occurrence of defects. In the literature, this group of methods is called "novel".

3. Novel Mode Shape-Based Damage Detection Methods

3.1. *Irregularity-based damage detection method*

Wang[17] and Wang Qiao,[18] respectively, introduced and developed a method to detect defects based on irregularities in mode shapes that might be caused by damage. The proposed method assumes that the measured mode shapes of the damaged structure can be broken down into a regular component (smooth) and an irregular component, and it consists of measuring noise and irregularities in the mode shape caused by the occurrence of damage. Only the irregular portion contains information about the damage, but it is usually invisible in the measured signal, as the regular component assumes much greater value, making it difficult to identify the damage. To overcome this problem, the irregular part of the signal is extracted. This process begins with signal filtering

$$r(x) = (\varphi \otimes h)(x) = \int_{-\infty}^{+\infty} \varphi(x - \tau) \cdot h(\tau)d\tau \qquad (1)$$

where $\varphi(x)$ is the displacement mode shape, $r(x)$ is a smoothed mode shape, $h(\tau)$ is a function used for smoothing and symbol $\otimes$ denotes convolution operation. Subsequently, the resulting regular component is subtracted from the original form of vibration, resulting in an irregular component $ir(x)$

$$ir(x) = \varphi(x) - r(x) \qquad (2)$$

The authors[18] used the two types of low-pass filters. The first was a Gaussian filter whose weight function is defined as

$$h(\tau) = \frac{1}{\alpha\lambda_c}e^{-\pi\left(\frac{\tau}{\alpha\lambda_c}\right)^2} \qquad (3)$$

where $\alpha = \sqrt{\ln 2/\pi}$ and λ_c is the cut-off wavelength. A second filter is a triangular weighting function which is expressed as

$$h(\tau) = \frac{1}{B} - \left(\frac{1}{B}\right)^2 |\tau| \tag{4}$$

where B is the filter cut-off length.

The effectiveness of this method for the detection of defects was verified in Ref. 18, in which the results were obtained for both the model and the real measurements of notched cantilever beam mode shapes. The results obtained revealed that the proposed method is suitable for the detection of both single and multiple damages in beams.

3.2. *Modified Laplace operator*

One of the first methods for damage detection that uses a single set of measured MSs for damage detection was proposed by Ratcliffe[19] and was called the modified Laplacian operator (MLO).

It is assumed that the MS of a healthy structure is smooth and can be approximated by a polynomial. When damage occurs, a singularity in this function appears, for instance, due to a local reduction in bending stiffness. To extract information about the position of this peculiar region, the difference between the Laplacian of MS and the polynomial that is approximating this Laplacian is determined. Polynomial coefficients are estimated for each Laplacian point separately based on two neighboring points on each side of the examined one. A graphical representation of estimation of a MLO index value at point i is presented in Figure 1.

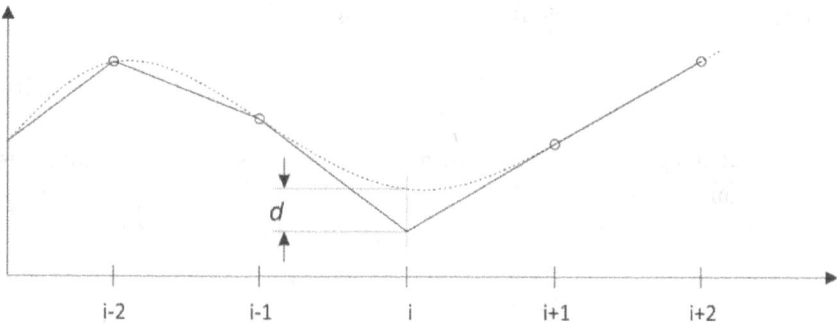

Figure 1. Difference between Laplacian (solid line) and its approximation (dotted line) at point i.

The MSC for a 1D structure such as a rod or a beam is expressed as

$$\varphi''(x) = \nabla^2 \varphi(x) = -\frac{M}{EI(x)} \tag{5}$$

where $\varphi''(x)$ denotes the MSC, M is the bending moment, and $EI(x)$ is bending stiffness.

The discrete Laplace operator that is most commonly used in structural health monitoring (SHM) utilizes a second-order central difference algorithm that can be written as

$$\nabla^2 \varphi(x) \approx \frac{\varphi(x - h_x) - 2\varphi(x) + \varphi(x + h_x)}{h_x^2} \tag{6}$$

where ∇^2 is the Laplace operator and h_x denotes the sampling interval. The MLO was further extended for 2D MSs by Qiau et al.,[20] where the difference between a 2D MSC and its approximation was used. For structures like shells or plates, the 2D MSC is defined as

$$\nabla^2 \varphi(x, y) = \nabla^2 \varphi(x) + \nabla^2 \varphi(y) = \frac{\partial^2 \varphi(x, y)}{\partial x^2} + \frac{\partial^2 \varphi(x, y)}{\partial y^2} \tag{7}$$

where

$$\frac{\partial^2 \varphi(x, y)}{\partial x^2} = -\frac{M_x(x, y)}{D} \tag{8}$$

$$\frac{\partial^2 \varphi(x, y)}{\partial y^2} = -\frac{M_y(x, y)}{D} \tag{9}$$

$M_x(x, y)$, $M_y(x, y)$ are bending moments in the x and y directions, respectively, and D is flexural rigidity expressed as

$$D = \frac{Eh^3}{12(1 - v^2)} \tag{10}$$

where E denotes Young's modulus, h is the thickness of the plate and v is the Poisson's ratio.

The 2D discrete central difference algorithm may be expressed as

$$\nabla \varphi(x) + \nabla \varphi(y) \approx \frac{\varphi(x - h_x) - 2\varphi(x) + \varphi(x + h_x)}{h_x^2}$$

$$+ \frac{\varphi(y - h_y) - 2\varphi(y) + \varphi(y + h_y)}{h_y^2} \tag{11}$$

where h_x and h_y denote sampling intervals in the x and y directions, respectively. In the literature, this method is also known as the gapped smooth method, wherein the squared value of the difference between the Laplacian operator and its polynomial approximations is used.

The Laplace operator is very effective in extracting any singularities in MS that can be caused by defects. However, this operator also intensifies the noise in the signal. To overcome this problem, Cao and Qiao[21] proposed to use a revised Laplace operator called *á trous* Laplace operator. This name comes from the wavelet transform (WT) algorithm *á trous* proposed by Shensa.[22] The new approach allowed determination of the MSC with various resolutions, increasing the likelihood of extracting information about the singularity of the signal containing interference. The mask of the proposed Laplacian takes the form

$$l_n = [1, \Theta_n, -2, \Theta_n, 1], \quad n \in \mathbb{N} \qquad (12)$$

where Θ_n is the vector containing $n - 1$ zeros, n is natural number defining the size of proposed Laplacian mask containing $2n + 1$ elements (e.g. $l_3 = [1, 0, 0, -2, 0, 0, 1]$).

Treating MS as a discrete function, the proposed *á trous* Laplace operator is calculated as a convolution

$$\varphi_n'' = \nabla^2 \varphi_n = \varphi * l_n \qquad (13)$$

In this way, the *á trous* Laplace operator produces a set of MSCs with different observation resolutions. Hence, this provides a more versatile tool where different sizes of MS features can be observed at different operator scales.

The *á trous* Laplace operator can be further enhanced by using a low-pass Gaussian filter that reduces the noise level in the signal.

$$\varphi_{n,\sigma}'' = \nabla^2 \varphi_{n,\sigma} = (\varphi * g_\sigma) * l_n = \varphi * (g_\sigma * l_n) \qquad (14)$$

where g_σ is the Gaussian filter mask with σ standard deviation.

3.3. *Spectral modal curvature*

As mentioned already, use of the central difference method for estimation of the MSC (modal curvature) is reported extensively in the literature. However, it has a major drawback. Even small measurement noise can produce large error values in derivatives calculated by this technique.

Yang *et al.* proposed to use modal curvature estimation by the Fourier spectral method. This technique can be applied for both 1D[23] and 2D mode shapes.[24]

For the 2D MS $\varphi(x, y)$, the Fourier transform (FT) is given by

$$\hat{\varphi}(k_x, y) = \int_{-\infty}^{+\infty} e^{-ik_x x} \varphi(x, y) dx \tag{15}$$

$$\hat{\varphi}(x, k_y) = \int_{-\infty}^{+\infty} e^{-ik_y y} \varphi(x, y) dy \tag{16}$$

The inverse FT (IFT)

$$\varphi(x, y) = \frac{1}{2\pi} \int_{-\infty}^{+\infty} e^{ik_x x} \hat{\varphi}(k_x, y) dk_x \tag{17}$$

$$\varphi(x, y) = \frac{1}{2\pi} \int_{-\infty}^{+\infty} e^{ik_y y} \hat{\varphi}(x, k_y) dk_y \tag{18}$$

where k_x and k_y are wavenumbers.

Due to the property that the FT of the n-order derivative can be expressed as

$$\hat{f^{(n)}}(\xi) = (2\pi i \xi)^n \cdot \hat{f}(\xi) \tag{19}$$

the MSC can be derived as

$$\varphi''(x, y) = -\frac{1}{2\pi} \int_{-\infty}^{+\infty} e^{ik_x x} k_x^2 \hat{\varphi}(k_x, y) dk_x$$
$$-\frac{1}{2\pi} \int_{-\infty}^{+\infty} e^{ik_y y} k_y^2 \hat{\varphi}(x, k_y) dk_y \tag{20}$$

The measured MS is a discrete function that can be defined as a function in interval $[0, 2\pi]$. The discrete FT (DFT) may be used for its curvature estimation as

$$\hat{\varphi}_{k_x, y} = \Delta x \sum_{k_x} e^{-ik_x x} \varphi_{x, y} \tag{21}$$

$$\hat{\varphi}_{x, k_y} = \Delta y \sum_{k_y} e^{-ik_y y} \varphi_{x, y} \tag{22}$$

The inverse algorithm for the inverse DFT (the IDFT) is given by

$$\varphi_{x,y} = \frac{1}{2\pi} \sum_{k_x} e^{ik_x x} \hat{\varphi}_{k_x,y} \tag{23}$$

$$\varphi_{x,y} = \frac{1}{2\pi} \sum_{k_y} e^{ik_y y} \hat{\varphi}_{x,k_y} \tag{24}$$

Finally, the discrete MSC in terms of the DFT can be derived by the equation

$$\varphi''_{x,y} = -\frac{1}{2\pi} \sum_{k_x} k_x^2 e^{ik_x x} \hat{\varphi}_{k_x,y} - \frac{1}{2\pi} \sum_{k_y} k_y^2 e^{ik_y y} \hat{\varphi}_{x,k_y} \tag{25}$$

Yang also proposed to use wavenumber domain filtering.[24,25] In the wavenumber domain, a low-pass filter can be used to reduce noise without diminution of the information about presumable MS singularities introduced by damage in the tested structure.

3.4. Simplified gapped smooth method

A similar but simplified algorithm of the MLO, called the simplified gapped smooth method, was proposed by Qiao and Wang.[26] The basis of this approach lies in the determination of a polynomial approximating a function that describes the deflection of the structure under the unified load surface (ULS). The ULS is defined as

$$ULS = F \cdot L \tag{26}$$

where $L = \{1, \ldots, 1\}^T_{1xN_i}$ is a vector representing the uniform load over the entire length of the specimen and F is the flexibility matrix that can be obtained from the equation

$$[F] = [\varphi][\Omega]^{-1}[\varphi]^T = \sum_{m=1}^{N_m} \frac{1}{\omega_m^2} \{\varphi_m\}\{\varphi_m\}^T \tag{27}$$

where $[\varphi] = [\{\varphi_1\}, \{\varphi_2\}, \ldots \{\varphi_{N_m}\}]$ is the matrix of MSs, $\{\varphi_m\}$ is m^{th} order MS. The values that are on the diagonal of stiffness matrix $[\Omega]$ correspond to the value ω_m^2 where ω_m is the mth natural frequency. Each column of $[F]$ represents the distribution of displacements of the structure under the influence of a unit force applied at the point that is the number of the column.

Polynomial functions approximating the ULS can be written as

$$\mathrm{ULS}^{\mathrm{approx}}(x) \approx c_0 - c_1 x - c_2 x^2 - c_3 x^3 - c_4 x^4 \tag{28}$$

The coefficients $c_0 c_1 c_2 c_3$ and c_4 are determined by regression analysis of the inspected ULS functions.

The deviation of a polynomial $\mathrm{ULS}^{\mathrm{approx}}$ from function ULS analysis allows one to determine the location of damage. For all available MSs, SGSM takes the form

$$\mathrm{SGSM}_i = \sum_{m=1}^{N_m} (\mathrm{ULS}_{i,m} - \mathrm{ULS}_{i,m}^{\mathrm{approx}})^2 \tag{29}$$

where i is the measuring point number, m is the number of MSs, and N_m is the number of investigated MSs.

3.5. Wavelets

The WT can be considered as an extension of the FT of variable size and position of windows. The advantage of the WT is that it allows analysis of the local signal with a different resolution of time-frequency (space-wavelength). The main advantage of applying the WT to damage detection is that it has the ability to detect singularities of the analyzed function. Because structural defects introduce into a system disruption of its dynamics, a large wavelet value can be used to identify the location and severity of damage.

The WT is a function used to decompose a signal $f(x)$ into a number of components $\psi_{u,s}(x)$ obtained from the mother wavelet $\psi(x)$ by a scaling operation (parameter s) and translational operation (parameter u) expressed by the equation

$$\psi_{u,s}(x) = \frac{1}{\sqrt{s}} \psi\left(\frac{x-u}{s}\right) \tag{30}$$

Considering the MS, $\varphi(x)$, as a 1D spatial signal, a continuous WT is given as[28,29]

$$W\varphi_{u,s}(x) = \frac{1}{\sqrt{s}} \int_{-\infty}^{+\infty} \varphi(x)\psi\left(\frac{x-u}{s}\right) dx \tag{31}$$

Any abrupt change in the value or impulse in the coefficient of wavelet can be used as an indicator of the location of damage.

For measurements made on the surface of a specimen, the WT takes the 2D form

$$W\varphi_{u,v,s}(x,y) = \frac{1}{s} \int_{-\infty}^{+\infty} \int_{-\infty}^{+\infty} \varphi(x,y)\psi\left(\frac{x-u}{s}, \frac{y-v}{s}\right) dxdy \qquad (32)$$

WT-based methods are a fast-growing group in recent years. The wavelet produced by the nth order differentiation has the nth order vanishing moments.

Rucka and Wilde[28] confirmed the effectiveness of the continuous WT (CWT) in damage detection using a fundamental MS, a beam-like structure, and a plate-like structure for both numerical and experimental data. Han *et al.*[30] proposed using the wavelet packet transform for the analysis of vibrational signals, defining the factor of damage by the wavelet packet energy rate index.

One of the most commonly used wavelets in SHM is the Gaussian wavelet. Let us take the Gaussian function g given by

$$g(x|\mu,\sigma) = \frac{1}{\sqrt{2\sigma^2\pi}}e^{-\frac{(x-\mu)^2}{2\sigma^2}} \qquad (33)$$

and assume its "standard" form g^0, which has standard deviation $\sigma = 1/\sqrt{2}$ and mean value $\mu = 0$, that can be expressed as

$$g^0(x) = \frac{1}{\sqrt{\pi}}e^{-x^2} \qquad (34)$$

The Gaussian function $g^0(x)$ can be used as a base for the mother wavelet as

$$g^n(x) = C_n(-1)^{n+1}\frac{\partial^n g^0(x)}{\partial x^n}, \quad n > 0 \qquad (35)$$

where C_n is a normalization constant, n is the number of vanishing moments. From $g^n(x)$, a family of Gaussian wavelets can be obtained as

$$g_{u,s}^n(x,y) = \frac{1}{\sqrt{s}}g^n\left(\frac{x-u}{s}\right) \qquad (36)$$

In an analogous way to Equations (33)–(36), a family of 2D Gaussian wavelets can be derived as

$$g^0(x, y) = \frac{1}{\sqrt{\frac{\pi}{2}}} e^{-(x^2+y^2)} \tag{37}$$

$$g^{m,n}(x, y) = C_{m,n}(-1)^{m+n} \frac{\partial^{m+n} g^0(x, y)}{\partial x^m \partial y^n} \tag{38}$$

$$g_{u,v,s}^{m,n}(x, y) = \frac{1}{s} g^{m,n}\left(\frac{x-u}{s}, \frac{y-v}{s}\right), \quad m, n > 0 \tag{39}$$

where m and n are orders of the wavelets (number of vanishing moments) in the x and y directions, respectively.

3.6. Directional wavelets

Xu et al.[31] proposed to enhance the 2D WT by adding a rotational parameter. This provides a 2D directional wavelet that can be derived from the equation

$$\psi_{u,v,s}(x, y) = \frac{1}{s} \psi\left(\frac{x'-u}{s}, \frac{y'-v}{s}\right) \tag{40}$$

where

$$\begin{Bmatrix} x' \\ y' \end{Bmatrix} = \begin{pmatrix} \cos\theta & \sin\theta \\ -\sin\theta & \cos\theta \end{pmatrix} \begin{Bmatrix} x \\ y \end{Bmatrix} \tag{41}$$

and θ is the orientation of the wavelet.

For a 2D MS, $\varphi(x, y)$, the directional WT is given as

$$W\varphi_{u,v,s,\theta}(x, y) = \frac{1}{s} \int_{-\infty}^{+\infty} \int_{-\infty}^{+\infty} \varphi(x, y)\psi$$

$$\times \left(\frac{x\cos\theta + y\sin\theta - u}{s}, \frac{-x\sin\theta + y\sin\theta - v}{s}\right) dx dy \tag{42}$$

In Ref. 31, directional Gaussian wavelets with $m = 2$ and $n = 2$ were used. A family of such directional wavelets is given by

$$\psi_{u,v,s,\theta}^{2,2}(x,y) = e^{-\left(\frac{x\cos\theta + y\sin\theta - u}{s}\right)^2 - \left(\frac{-x\sin\theta + y\sin\theta - v}{s}\right)^2}$$

$$\cdot \frac{1}{3s\sqrt{\frac{\pi}{2}}} \left(-2 + 4\left(\frac{x\cos\theta + y\sin\theta - u}{s}\right)^2 \right.$$

$$\left. -2 + 4\left(\frac{-x\sin\theta + y\sin\theta - v}{s}\right)^2 \right) \tag{43}$$

This wavelet ($\psi_{u,v,s,\theta}^{2,2}(x,y)$) has a number of mathematical properties, such as smoothness, symmetry, differentiability, localizability, and directionality, which make it an excellent tool for damage detection.

3.7. *Wavelet mode shape curvature*

Another approach for MSC estimation without using second-order central difference was proposed by Cao *et al.*[32] In order to smoothen noise from an MSC φ'', a Gaussian function g may be applied. Taking into account the convolution ($\otimes$) property of differentiation, the following relation may be written:

$$\varphi'' \otimes g = \varphi \otimes g'' \tag{44}$$

where g'' is the second-order derivative of the Gaussian function, known as the Mexican hat wavelet with two vanishing moments.[33] Incorporating translation u and scale s parameters into 1D the Gaussian function, $g(x)$, a $g_{u,s}(x)$ is given by

$$g_{u,s}(x) = \frac{1}{\sqrt{s}} g\left(\frac{x - u}{s}\right) \tag{45}$$

The mother wavelet from Equation (45) is consecutively used for creating a family of wavelets as

$$g_{u,s}''(x) = s^2 \frac{\partial^2}{\partial x^2} g_{u,s}(x) = \frac{1}{\sqrt{s}} g''\left(\frac{x - u}{s}\right) \tag{46}$$

Substituting the scaled g and g'' gives the equation for a multiscale estimation of the MSC

$$\varphi'' \otimes g_{u,s} = \varphi \otimes g_{u,s}'' \tag{47}$$

Let $g''_{u,s}$ be represented by $g''_s(u)$ for notational convenience and considering symmetry of the Gaussian function and its derivative, a final formulation for the wavelet mode shape curvature (WT MSC) $\varphi^*_s(u)$ is given as

$$\varphi^*_s(u) = \varphi \otimes g''_s(u) = \frac{1}{\sqrt{s}} \int_{-\infty}^{+\infty} \varphi(x) g'' \left(\frac{x-u}{s} \right) dx$$

$$= s^{\frac{3}{2}} \varphi \otimes \left[\frac{d}{dx^2} g \left(\frac{x}{s} \right) \right] (u)$$

$$= s^{\frac{3}{2}} \frac{d}{dx^2} \left[\varphi \otimes g \left(\frac{x}{s} \right) \right] (u)$$

$$= s^2 \frac{d}{dx^2} \left[\varphi \otimes g_s(u) \right] \tag{48}$$

The final form of Equation (48) is composed of two important operations: convolution of the MS φ and the scaled Gaussian function $g_s(u)$, and second-order differentiation of this convolution result. A scaling operation (product with s^2) does not change the result from the damage identification viewpoint. The same solution can be extended for 2D mode shapes.[34]

The 2D Mexican hat wavelet is given as

$$g''_{u,v,s}(x,y) = s^2 \left(\frac{\partial^2}{\partial x^2} + \frac{\partial^2}{\partial y^2} \right) g_{u,v,s}(x,y)$$

$$= s \left(\frac{\partial^2}{\partial x^2} + \frac{\partial^2}{\partial y^2} \right) g \left(\frac{x-u}{s}, \frac{y-v}{s} \right)$$

$$= \frac{1}{s} g'' \left(\frac{x-u}{s}, \frac{y-v}{s} \right) \tag{49}$$

where $g_{u,v,s}$ is the scaled and translated 2D Gaussian function

$$g_{u,v,s}(x,y) = \frac{1}{s} g \left(\frac{x-u}{s}, \frac{y-v}{s} \right) \tag{50}$$

u and v are translation parameters in the x and y directions, respectively.

Therefore, the 2D WT MSC takes the form

$$\varphi_s^*(u, v) = \varphi \otimes g_s''(u, v)$$

$$= \frac{1}{s} \int_{-\infty}^{+\infty} \int_{-\infty}^{+\infty} \varphi(x, y) g'' \left(\frac{x-u}{s}, \frac{y-v}{s} \right) dx dy$$

$$= s \left(\frac{\partial^2}{\partial x^2} + \frac{\partial^2}{\partial y^2} \right) [\varphi \otimes g_s(u, v)] \tag{51}$$

3.8. *Teager energy of wavelet mode shape curvature*

As stated already, the modal curvature is an excellent tool for exaggerating singularities in MS and thus may be used for damage localization. Local changes in bending stiffness create a singularity in modal curvature with a physical implication for damage detection and assessment. However, the global fluctuation trend of modal curvature can obscure indications of damage. To overcome this problem, the Teager energy operator (TEO) may be used, as proposed by Xu *et al.*[35]

TEO is a nonlinear operator for calculating the energy of a signal in a point-wise manner. It was originally developed[36] for analysis of sound signals. The TEO for a 1D signal is defined as

$$\Psi(x(t)) = (x')^2(t) - x(t)x''(t) \tag{52}$$

for continuous signals. Kaiser proposed its discrete version[37] (also called the Teager–Kaiser Operator) which is defined as

$$\Psi[x(n)] = x^2[n] - x[n-1]x[n+1] \tag{53}$$

This operator has the capability of amplifying local singularities of the signal from a nonlinear viewpoint and has the interesting property

$$\Psi[x(n) + C] = \Psi[x(n)] - Cx''(n) \tag{54}$$

where C is a constant value.

For discretizing WT MSC, $\varphi_s^*(u)$ to $\varphi_s^*[n]$, the 1D TEO–WT MSC can be expressed as

$$\Psi[\varphi_s^*[n]] = (\varphi_s^*[n])^2 - \varphi_s^*[n-1]\varphi_s^*[n+1] \tag{55}$$

Combining these two approaches gives a DI retaining its physical meaning. It is robust against noise and it brings out small singularities, removing the global trend at the same time.

The same approach can easily be extended to plate-like structures. The 2D discrete TEO–WT MSC takes the form[34]

$$\Psi[\varphi_s^*[j,k]] = 2(\varphi_s^*[j,k])^2 - \varphi_s^*[j-1,k]\varphi_s^*[j+1,k] - \varphi_s^*[j,k-1]\varphi_s^*[j,k+1] \tag{56}$$

where $\varphi_s^*[j,k]$ is discrete versions of $\varphi_s^*(u,v)$

3.9. Multiscale shear strain gradient

Delamination is a very specific type of damage that occurs in composite plates. In most cases, it takes the form of separation between layers of laminate due to imperfect bonding, local impact, or thermal loading. A special tool proposed by Cao et al.[38] for detection of this type of damage based on mode shape measurements is called the multiscale shear strain gradient (MSG).

Using Kirchhoff plate theory, a physical background for this technique can be given. The strain components of a laminate in the plane-stress state, when the middle of the plate coincides with the x–y plane, are given by the equations

$$\varepsilon_{xx} = -z\frac{\partial^2 w(x,y)}{\partial x^2} \tag{57}$$

$$\varepsilon_{yy} = -z\frac{\partial^2 w(x,y)}{\partial y^2} \tag{58}$$

$$\varepsilon_{xy} = -2z\frac{\partial^2 w(x,y)}{\partial x \partial y} \tag{59}$$

where $\varepsilon_{xx}, \varepsilon_{yy}$ denote normal strains parallel to the x and y axes, respectively, and ε_{xy} is the shear strain in the x–y plane. Changes in any of these strains can potentially be used to detect and assess damage. However, for delamination that is mostly separation of neighboring material layers, most of the strain changes appear in ε_{xy}. Due to the fact that small delamination can cause local but very insignificant changes in shear strain, a gradient operator that functions cumulatively in both x and y directions can be applied to extract this information. This operation is expressed as

$$\vec{\varepsilon}_{xy} = -\frac{\partial^2 \varepsilon_{xy}}{\partial x \partial y} = -2z\frac{\partial^4 w(x,y)}{\partial x^2 \partial y^2} \tag{60}$$

where $\vec{\varepsilon}_{xy}$ is the shear strain gradient. The differentiation operation amplifies any local changes in the shear strain of an inspected structure. At the same time, unfortunately, noise from the MS function is also increased. This can overshadow any indication of a small delamination. To avoid this disadvantage, a multiscale transformation is used by convoluting $\vec{\varepsilon}_{xy}$ with the scaled Gaussian function g_s (Equation (50)), giving

$$\vec{\varepsilon}_{xy}^{s} = \vec{\varepsilon}_{xy} \otimes g_s = -2z\frac{\partial^4 g_s}{\partial x^2 \partial y^2} \otimes w \tag{61}$$

where $\vec{\varepsilon}_{xy}^{s}$ is the MSG and w is the transverse deflection of the middle surface of the plate-like structure.

3.10. *Fractal dimension*

The Fractal dimension (FD) was originally proposed by Mandelbrot.[39] In 1988, Katz defined an approximation method of FD curvature using a sequence of points

$$\mathrm{FD}_m(x) = \frac{\log(n)}{\log(n) + \log\left(\frac{d(x_i,M)}{L(x_i,M)}\right)} \tag{62}$$

where

$$L(x_i, M) = \sum_{k=1}^{M} \sqrt{(y(x_{i+k}) - y(x_{i+k-1}))^2 + (x_{i+k} - x_{i+k-1})^2} \tag{63}$$

$$d(x_i, M) = \max_{1 \leq k \leq M} \sqrt{(y(x_{i+k}) - y(x_i))^2 + (x_{i+k} - x_i)^2} \tag{64}$$

where M is the size of the window sliding along the tested function.

However, this method does not work properly for higher order of MSs, giving false peaks in the local maxima and minima of the MSs' first derivatives. To overcome this problem, Wang and Qiao[26] proposed a scaled version for calculating the FD by adding the scale factor s. This method is called a generalized FD (GFD) and can be defined as

$$\text{GFD}_m(x) = \frac{\log(n)}{\log(n) + \log\left(\frac{ds(x_i,M)}{Ls(x_i,M)}\right)} \tag{65}$$

where

$$L(x_i, M) = \sum_{k=1}^{M} \sqrt{(y(x_{i+k}) - y(x_{i+k-1}))^2 + s^2(x_{i+k} - x_{i+k-1})^2} \tag{66}$$

$$d(x_i, M) = \max_{1 \le k \le M} \sqrt{(y(x_{i+k}) - y(x_i))^2 + s^2(x_{i+k} - x_i)^2} \tag{67}$$

Hadjileontiadis and Douka[40] extended the GFD method for 2D objects. They confirmed its effectiveness and high noise immunity based on the results obtained from the numerical model. For simplicity of notation, let $\Im$ be the operator defined by Equation (65). Accordingly, a 2D operator is applied to horizontal, vertical, and diagonal dimensional slices of 2D MSs $\varphi(x, y)$, resulting in corresponding arrays of fractal dimensions FD^H, FD^V, and FD^D size LxL and consisting of

$$FD^H(i) = \Im\{\varphi(i, 1:L)\}, \quad i = 1, \ldots, L \tag{68}$$

$$FD^V(i) = \Im\{\varphi(1:L, i)\}, \quad i = 1, \ldots, L \tag{69}$$

$$FD^D(i) = \begin{cases} \Im\{\varphi(i+1:L, i:L-1)\}, \\ \Im\{\varphi(1:L, 1:L)\}, \\ \Im\{\varphi(1:L-1, i+1:L)\}, \end{cases} \quad i = 1, \ldots, L-1 \tag{70}$$

The determined matrices FD^H, FD^V, and FD^D are used for fault localization in plate-like structures. Depending on which array indicates the occurrence of damage, its orientation can be assessed.

3.10.1. *Affine transformation*

For better performance of the FD, Bai[27] proposed, using an affine transformation in preprocessing MSs, to remove local extrema of the FD. This operation can be expressed as

$$\begin{Bmatrix} x'_i \\ y'_i \end{Bmatrix} = A \begin{Bmatrix} x^*_i \\ y^*_i \end{Bmatrix}, \quad A = \begin{bmatrix} 1 & 0 \\ \sin\theta & (\cos\theta)/k \end{bmatrix} \tag{71}$$

where A is the affine transformation matrix, k and θ are adjustable parameters of this transformation. Those parameters can be chosen from a wide range of values giving transformed MSs different degrees of smoothness but having no local extreme, which is the aim of this operation.

3.11. *Wavelet mode shape curvature fractal analysis*

In the previous sections, WT-based methods as well as FD-based methods for detecting damage were presented. The latter is an excellent tool for multiresolution insight into signals and the former is a very effective tool for detecting singularities of the analyzed function. To use the advantages of both approaches, two signal processing algorithms can be combined. The name for this is wavelet-aided fractal analysis.[42]

3.11.1. *Scale MS*

The measured MS for an inspected structure can be divided into three main components: noise, singularities caused by damage, and the main part of the MS that is its exact component. To gain insight into those components separately, the scale MS (SMS) has been proposed. Wavelets have the capability of performing analysis of 2D signals in multiresolution manner. To create a different scale of MS, a 2D wavelet (e.g. 2D Gabor wavelet,[41] Gaussian wavelet[42]) is used to perform a transformation represented by a convolution regime as

$$W^s : (x_{i,j}, x_{i,j}, w_{i,j}^s) = W : (x_{i,j}, x_{i,j}, w_{i,j}) \otimes \psi_s(x, y) \qquad (72)$$

where $W^s : (x_{i,j}, x_{i,j}, w_{i,j}^s)$ is a scaled MS at scale s and $\otimes$ denotes convolution.

A proper scale is chosen such that it carries information about possible damage and at the same time most of the noise and general MS trend information are not conveyed.

3.11.2. *Waveform fractal analysis*

In a second step, a chosen SMS in divided into horizontal and vertical sets of lines as

$$W^s : (x_{i,j}, x_{i,j}, w_{i,j}^s) \approx W_x^s : (x_{i,j}, y_j, w_{i,j}^s) + W_y^s : (x_i, y_{i,j}, w_{i,j}^s) \qquad (73)$$

where $W_x^s : (x_{i,j}, x_{i,j}, w_{i,j}^s)$ and $W_y^s : (x_{i,j}, x_{i,j}, w_{i,j}^s)$ are SMS lines in the x and y directions respectively. By performing fractal analysis (see point 3.10) through all set of lines, damage detection is performed.

3.11.3. Fractal complexity

Another approach was proposed by Cao,[41] in which an SMS obtained in the same way is analyzed using fractal complexity based on the Kolmogorov capacity dimension.[43] Assume that a set Ξ is covered by a number of same-sized hypercubes with the side dimension equal to r. The minimum number of hypercubes that will cover the whole Ξ is denoted as (r). When r approaches zero, the capacity dimension $D_c(\Xi)$ can be derived from the equation

$$D_c(\Xi) = \lim_{r \to 0} \frac{\log N(r)}{\log \frac{1}{r}} \tag{74}$$

Capacity dimension analysis is performed at the chosen scale of the SMS using a moving window that creates the set $\Xi_{i,j}$ having a central point at $(x_i y_i)$. In this way, a complexity value at each point of the SMS is determined and a map of this value can be determined. Local changes in the complexity function may be used as damage indicators.

3.12. Summary

The presented techniques are based on processing MSs or ODSs. They have physical implications and at the same time do not require any reference state or material properties of the examined structure.

The choice of the appropriate method should depend on the type of the examined structure, potential type of damage that may occur, computing power of the used equipment and the experience of the person conducting the tests and calculations.

It is a good practice to use several methods to ensure maximum damage detection efficiency and eliminate possible false alarms. This approach also provides a possibility of data fusion for different DIs to create one hybrid indicator as shown in the next section.

4. Data Fusion

Some region of particular mode shape may be insensible for damage (e.g. crack occurring near the node of a MS). Also, false damage detection may occur due to unideal measurement conditions. In order to accommodate this deficiency, a group of MSs obtained at different frequencies should be tested. To fuse data, a standard score normalization procedure is applied to a DI as

$$\chi' = \frac{\chi - \mu(\chi)}{\sigma(\chi)} \tag{75}$$

where χ is a DI such as FD, WT–TEO, etc., assigned for a particular MS, $\mu(\chi)$ and $\sigma(\chi)$ are the mean and standard deviation of χ and χ' denotes a standardized DI χ.

This normalization process has two advantages:

(1) It provides a proper scaling for the DI determined for various MSs as well as allows for fusion of number of DIs determined for the same MS,
(2) noisy DIs have lower weight due to higher mean value and standard deviation, therefore lower values after normalization contributes less to the FDI.

The rescaling operation does not result in damage characterization shading, but allows for multiple indexes to be fused with adequate weight

$$\bar{\chi} = \frac{1}{NM} \sum_{i=1}^{N} \sum_{j=1}^{M} \chi'_{i,j} \tag{76}$$

where $\bar{\chi}$ is the FDI including N number of mode shapes and M is the number of various DIs.

References

1. Dossing, O. Structural stroboscopy — measurement of operational deflection shapes. *Sound and Vibration Magazine* **1**, pp. 110–116 (1988).
2. Richardson, M. H. Is it a mode shape, or an operating deflection shape? *Sound and Vibration Magazine, 30th Annual Issue* 1, pp. 1–10 (1997).
3. West, W. M. Illustration of the use of modal assurance criterion to detect structural changes in an orbiter test specimen. In *Proceedings of the Air Force C Aircraft Structural Integration*, pp. 1–6 (1984).
4. Allemang, R. J. and Brown, D. L. A correlation coefficient for modal vector analysis. In *Proceedings of IMAC*, Vol. 1, pp. 110–116 (in USA) (1982).
5. Lieven, N. A. J. and Ewins, D. J. Spatial correlation of mode shapes the coordinate modal assurance criterion (COMAC). In *Proceedings of IMAC*, Vol. 4, pp. 690–695 (in USA) (1988).

6. Pandey, A. K., Biswas, M., and Samman, M. M. Damage detection from changes in curvature mode shapes. *Journal of Sound and Vibration* **145**, pp. 321–332 (1991).

7. Qiao, P., Lestari, Shah, W. M. G., and Wang, J. Dynamics-based damage detection of composite laminated beams using contact and noncontact measurement systems. *Journal of Composite Materials* **10**(41), pp. 1217–1252 (2007).

8. Maia, N. M. M., Silva, J. M. M., and Almas, E. A. M. Damage detection in structures: From mode shape to frequency response function methods. *Mechanical Systems and Signal Processing* **17**(3), pp. 489–498 (2003).

9. Ho, Y. K. and Ewins, D. J. On the structural damage identification with mode shapes. In *Proceedings of the International Conference on System Identification and Structural Health Monitoring*, pp. 677–684 (in Spain) (2000).

10. Cornwell, P. Application of the strain energy damage detection method to plate-like structures. *Journal of Sound and Vibration* **224**(2), pp. 359–374 (1999).

11. Stubbs, N., Kim, J. T., and Farrar, C. R. Field verification of a nondestructive damage localization and severity estimation algorithm. In *Proceedings of SPIE — The International Society for Optical Engineering*, Vol. 2460, pp. 210–218 (in USA) (1995).

12. Choi, F. C., Li, J, Samali, B., and Crews, K. Application of the modified damage index method to timber beams. *Engineering Structures* **30**, pp. 1124–1145 (2008).

13. Sazonov, E. and Klinkhachorn, P. Optimal spatial sampling interval for damage detection by curvature or strain energy mode shapes. *Journal of Sound and Vibration* **285**, pp. 783–801 (2005).

14. Pandey, A. K. and Biswas, M. Damage detection in structures using changes in flexibility. *Journal of Sound and Vibration* **169**(1), pp. 3–17 (1994).

15. Zhang, Z. and Aktan, A. E. The damage indices for constructed facilities. In *Proceedings of IMAC*, Vol. 13, pp. 1520–1529 (in USA) (1995).

16. Doebling, S. W., Farrar, R. C., and Prime, M. B. A summary review of vibration-based damage identification methods. *Shock and Vibration Digest* **2**(30), pp. 91–105 (1998).

17. Wang, J. Damage detection in beams by roughness analysis. In *Proceedings of SPIE — The International Society for Optical Engineering*, Vol. 6174, pp. 488–499 (in USA) (2006).

18. Wang, J. and Qiao, P. On irregularity-based damage detection method for cracked beams. *International Journal of Solids Structure* **45**, pp. 688–704 (2008).

19. Ratcliffe, C. P. Damage detection using modified Laplacian operator on mode shape data. *Journal of Sound and Vibration* **204**(3), pp. 505–517 (1997).

20. Qiao, P., Lu, K., Lestari, W., and Wang, J. Curvature mode shape-based damage detection in composite laminated plates. *Composite Structures* **80**, pp. 409–428 (2007).

21. Cao, M. and Qiao, P. Novel Laplacian scheme and multiresolution modal curvatures for structural damage identification. *Mechanical Systems Signal Processing* **23**, pp. 1223–1242 (2009).

22. Shensa, M. J. The discrete wavelet transform: Wedding the a trous and Mallat algorithms. *IEEE Transaction of Signal Processing* **40**(10), pp. 2464–2482 (1992).

23. Yang, Z.-B., Radzieński, M., Kudela, P., and Ostachowicz, W. Fourier spectral-based model curvature analysis and its application to damage detection in beams. *Mechanical Systems Signal Processing* **84**(Part A), pp. 763–781 (2017).

24. Yang, Z.-B., Radzieński, M., Kudela, P., and Ostachowicz, W. Two-dimensional modal curvature estimation via Fourier spectra method for damage detection. *Composite Structures* **158**, pp. 155–167 (2016).

25. Yang, Z.-B., Radzieński, M., Kudela, P., and Ostachowicz, W. Scale-wavenumber domain filtering method for curvature modal damage detection. *Composite Structures* **154**, pp. 396–409 (2016).

26. Wang, J. and Qiao, P. Improved damage detection for beam-type structures using a uniform load surface. *Structural Health Monitoring* **6**(2), pp. 99–110 (2007).

27. Bai, R., Song, X., Radzieński, M., Cao, M., Ostachowicz, W., and Wang, S. S. Crack location in beams by data fusion of fractal dimension features of laser-measured operating deflection shapes. *Smart Structural Systems* **13**(6), pp. 975–991 (2014).

28. Rucka, M. and Wilde, K. Application of continuous wavelet transform in vibration based damage detection method for beams and plates. *Journal of Sound Vibration* **297**, pp. 536–550 (2006).

29. Morlier, J., Bos, F., and Castera, P. Diagnosis of a portal frame using advanced signal processing of laser vibrometer data. *Journal of Sound Vibration* **297**, pp. 420–431 (2006).

30. Han, J.-G., Ren, W.-X., and Sun, Z.-S. Wavelet packet based damage identification of beam structures. *International Journal Solids Structures* **42**, pp. 6610–6627 (2005).

31. Xu, W., Radzieński, W., Ostachowicz, W., and Cao, M. Damage detection in plates using two-dimensional directional Gaussian wavelets and laser scanned operating deflection shapes. *Structural Health Monitoring* **2**(5–6), pp. 457–468 (2013).

32. Cao, M. S., Xu, W., Ostachowicz, W., and Su, Z. Damage identification for beams in noisy conditions based on Teager energy operator-wavelet transform modal curvature. *Journal of Sound and Vibration* **333**, pp. 1543–1553 (2014).

33. Mallat, S. *A Wavelet Tour of Signal Processing*, 3rd edn. Academic Press, San Diego (2008).

34. Xu, W., Cao, W., Ostachowicz, W., Radzieński, M., and Xi, N. Two-dimensional curvature mode shape method based on wavelets and Teager energy for damage detection in plates. *Journal of Sound and Vibration* **347**, pp. 266–278 (2015).

35. Xu, W., Cao, M., Radzieński, M., Xia, N., Su, Z., Ostachowicz, W., and Wang, S. S. Detecting multiple small-sized damage in beam-type structures by Teager energy of modal curvature shape. *JVE International Ltd. Journal of Vibroengineering* **17**(1), pp. 275–286 (2015).

36. Teager, H. M. and Teager, S. M. Evidence for nonlinear sound produc-
    tion mechanisms in the vocal tract, In eds. Hardcastle W. J. and Marchal
    A., *Speech Production and Speech Modelling.* NATO ASI Series (Series D:
    Behavioural and Social Sciences), **55**, pp. 241–261 (1990).
37. Kaiser, J. F. On a simple algorithm to calculate the "energy" of a signal. In
    Proceedings IEEE ICASSP- 90, pp. 381–384 (in USA) (1990).
38. Cao, M., Ostachowicz, W., Radzieński, M., and Xu, W. Multiscale shear-
    strain gradient for detecting delamination in composite laminates. *Applied
    Physics Letters* **103** (2013).
39. Mandelbrot, B. B. How long is the coast of Britain? Statistical self-similarity
    and fractional dimension. *Science* **156**, pp. 636–638 (1967).
40. Hadjileontiadis, L. J. and Douka, E. Crack detection in plates using fractal
    dimension. *Engineering Structures* **29**, pp. 1612–1625 (2007).
41. Cao, M., Xu, H., Bai, R., Ostachowicz, W., Radzieński, M., and Chen, L.
    Damage characterization in plates using singularity of scale mode shapes.
    Applied Physics Letters 106 (2015).
42. Bai, R., Radzieński, M., Cao, M., Ostachowicz, W., and Su, Z. Non-baseline
    identification of delamination in plates using wavelet-aided fractal analysis of
    two-dimensional mode shapes. *Journal International Material Systems and
    Structures* **26**(17), pp. 2338–2350 (2014).
43. Kolmogorov, A. N. A new metric invariant of transient dynamical systems
    and automorphisms in Lebesgue spaces. *Doklady Akademic Nauk SSSR* **124**,
    pp. 754–755 (1958).

Chapter 7

Damage Identification Based on Response Functions in Time and Frequency Domains

R. P. C. Sampaio[*,†,¶], T. A. N. Silva[‡,‖] N. M. M. Maia[†,**] and S. Zhong[§,††]

*Escola Naval, Alfeite, Almada 2810-001 Portugal

†LAETA, IDMEC, Instituto Superior Técnico, Departamento de Engenharia Mecânica, Universidade de Lisboa, AV. Rovisco Pais Lisboa 1049-001, Portugal

‡UNIDEMI, DEMI, Faculdade de Ciências e Tecnologia Universidade Nova de Lisboa, Caparica 2829-516, Portugal

§Laboratory of Optics, Terahertz and Non-Destructive Testing School of Mechanical Engineering and Automation, Fuzhou University Fuzhou 350108, P. R. China

¶chedas.sampaio@marinha.pt

‖tan.silva@fct.unl.pt

**nuno.manuel.maia@tecnico.ulisboa.pt

††zhongshuncong@hotmail.com

This chapter reviews and compares some of the most known methods of damage identification, in the frequency domain, which use operational deflection shapes (ODSs) built from the frequency response functions (FRFs). Two new methods of damage identification, in the time domain, using (ODSs) built from the time responses are also analyzed. To assess the performance of the different indicators, two numerical tests are undertaken with a finite element model (FEM) model, of a cantilever beam, to evaluate their detection and localizing ability.

Keywords: Damage identification; Time response; Frequency response; Mode shape analysis; Operational deflection shape (ODS); Modal assurance criterion (MAC); Coordinate modal assurance criterion (COMAC); Damage index (DI); Mode shape curvature (MSC); Sensitivity to damage analysis; Strain energy damage index (SEDI); Procedure of maximum occurrences (MO); Resonant gapped-smoothing broadband method (RGSB); Frequency domain assurance criterion (FDAC); Transmissibility damage indicator (TDI); Error in the constitutive relations (ECR); Time domain Katz fractal dimension (TKFD) method.

1. Introduction

The structural damage identification, also known as structural health monitoring (SHM), aims at answering the following questions[1,2,5]:

- Existence: Is there damage in the system?
- Localization: Where is the damage in the system?
- Type: What kind of damage is present?
- Extent: How severe is the damage?
- Prognosis: How much useful life remains?

The answers to the above questions define the two main objectives of the SHM:

- Detection (existence);
- Diagnostic (location, type, extent, prognosis).

The structural damage identification, based on measured vibration, assumes that any structural damage will change somehow one or more of the dynamic properties of the structure, like mass, damping, or stiffness, and consequently will change the vibration response.

The dynamic equilibrium of a structure with N degrees of freedom is often described by the following system of simultaneous equations[6,7]:

$$\mathbf{M}\ddot{\mathbf{x}} + \mathbf{C}\dot{\mathbf{x}} + \mathbf{K}\mathbf{x} = \mathbf{f}(\mathbf{t}) \tag{1}$$

where $\mathbf{M}$ is the mass matrix, $\mathbf{K}$ is the stiffness matrix, $\mathbf{C}$ is the viscous damping matrix, $\mathbf{x}(\mathbf{t})$ is the displacement vector, $\dot{\mathbf{x}}(\mathbf{t})$ is the velocity vector, $\ddot{\mathbf{x}}(\mathbf{t})$ is the acceleration vector, and $\mathbf{f}(\mathbf{t})$ is the excitation vector. The mass, damping, and stiffness matrices are elements of what is often called the *spatial model*.

Considering the homogenous solution of the system that models the free vibration, one obtains

$$\left(\mathbf{K} - \omega^2\mathbf{M} + i\omega\mathbf{C}\right)\psi = \mathbf{0} \tag{2}$$

which represents a generalized eigenvalue problem. The solution of this equation permits one to find the eigenvalues, ω_r^2 $(r = 1 \ldots N)$, and the eigenvectors, $\psi^{(r)}$, allowing the identification of the *modal model* constituted by the spectral matrix

$$\mathbf{\Omega} = \begin{bmatrix} \ddots & 0 & 0 \\ 0 & \bar{\omega}_r^2 & 0 \\ 0 & 0 & \ddots \end{bmatrix}$$

where $\bar{\omega}_r$ is the natural frequency of mode r, is the modal matrix $[\psi^{(1)}, \psi^{(2)} \dots \psi^{(r)} \dots \psi^{(N)}]$, where $\psi^{(r)}$ is the mode shape r, and the modal damping factors are denoted by ξ_r.

For the case of harmonic excitation, the relation between the response and the excitation, at each frequency of the analysis, is given by

$$\mathbf{X}(\omega) = \alpha(\omega) \mathbf{F}(\omega) \tag{3}$$

where

$$\alpha(\omega) = (\mathbf{K} - \omega^2 \mathbf{M} + i\omega \mathbf{C})^{-1} \tag{4}$$

is the system receptance matrix. If instead of the displacement one relates velocity against force or acceleration against force, $\alpha(\omega)$ is called mobility or accelerance matrix, respectively. Each element $\alpha_{i,j}(\omega)$ of the matrix corresponds to an individual frequency response function (FRF), describing the relation between the response at a particular coordinate i and a single force excitation applied at a coordinate j

$$\alpha_{i,j}(\omega) = \frac{X_i}{F_j}; \quad F_k = 0, \quad k = 1 \dots N; \quad k \neq j \tag{5}$$

The $\alpha(\omega)$ matrix constitutes what is called the *response model*. This model can also be represented by the responses, $\mathbf{x}(t)$ or $\mathbf{X}(\omega)$, at the measured, or calculated, coordinates i.

The column vectors of the system matrix, $\alpha(\omega)$, are the operational deflection shapes (ODSs) that describe the shape (in space) exhibited by the structure at each excitation frequency, ω, given by the responses normalized by the applied excitations. The ODS can also be the shape that the structure exhibits at each instant of time, t, given by the responses in time or the shape that the structure exhibits at each excitation frequency, ω, given by the responses in frequency. For the sake of clarity, the different nature of the ODSs will be distinguished by the nomenclature:

- time ODS, the shape at each instant of time;
- frequency ODS, the shape at each frequency;
- FRF-based ODS, the shape at each frequency calculated from the FRFs.

The methods of damage identification, based on measured vibration, are usually developed from the observed change of the models mentioned above, i.e. spatial, modal, and response models.

The methods discussed in this chapter, i.e. the methods of damage identification based on vibration response functions, in time and frequency domains, observe changes in the response model.

Even so, additional classification criteria are necessary, like whether to assume a linear behavior or not, whether to have a reference measurement or not, whether measurement of the excitation is needed or not, etc.

The methods presented in this chapter are, essentially, a logical generalization of the methods based on the modes of vibration in the sense that they use all the frequencies measured and not just the natural frequencies.

The methods based on changes in the spatial mode shapes, and their respective spatial derivatives, involve the calculation of correlations, absolute differences, weighted differences, differences normalized with the natural frequencies or norms and Ritz vectors, among others, which are then compared with numerical models (FEM, for example) or with reference measurements. Several indices, based on the orthogonality properties of the mode shapes, have been developed for the purpose of comparing measured mode shapes with numeric, or measured reference mode shapes, where the best known is the modal assurance criterion (MAC).[3,8]

Within these methods, one may highlight those based on the second spatial derivative of the mode shapes. The importance of these methods is that, in the case of beam-like structures, the curvature is directly related to the flexural rigidity[4]:

$$\frac{1}{\rho} = \frac{M}{EI} \tag{6}$$

where ρ is the radius of curvature, EI is the flexural rigidity (E, Young's modulus and I, second moment of area) and M is the moment applied to the beam, and the second derivative of displacement y being approximately equal to the curvature

$$\frac{1}{\rho} = \frac{\frac{d^2y}{dx^2}}{\left[1 + \left(\frac{dy}{dx}\right)^2\right]^{\frac{3}{2}}} \approx \frac{d^2y}{dx^2} \tag{7}$$

This way, it is possible to relate the second derivative of the mode, or curvature, with the flexural rigidity which varies with the damage. This derivative is usually calculated by finite differences, by splines or by analytic differentiation of interpolating polynomials.

In this chapter, the curvatures of modes and ODSs are calculated with the second-order finite central difference, respectively,

$$\psi''_{i,r} = \frac{\psi_{i-1,r} - 2\psi_{i,r} + \psi_{i+1,r}}{h^2} \tag{8}$$

and

$$\alpha''_{i,j}(\omega) = \frac{\alpha_{i-1,j}(\omega) - 2\alpha_{i,j}(\omega) + \alpha_{i+1,j}(\omega)}{h^2} \tag{9}$$

where h is the distance between adjacent coordinates. Considering h equal between all adjacent coordinates, one can consider it equal to 1.

Very similar to these methods are those based on the change of the strain modes shapes and respective spatial derivatives. The similitude comes from the fact that the curvature and the flexural strain, for beam-like structures, are directly related by

$$\varepsilon = \frac{y}{\rho} = \frac{M}{EI}y \approx \frac{d^2y}{dx^2}y \tag{10}$$

where ε is the strain. The strain can be directly measured by strain gauges or indirectly by the above relation. Several authors refer that strain, measured with strain gauges, is more sensitive to local damage than the displacement mode shapes and their derivatives.

The idea of identifying damage, based on the vibration response, comes from the need to surpass some problems of the existing methods based on the vibration modes, which are as follows[1]:

- near the resonances, the FRFs are less accurate;
- they are a particular form of data compression with obvious risks of losing useful information;
- time and expertise are required from the analyst in the modal identification process, which often leads to different results in the modal parameters when different analysts use the same data set;
- if there are excited natural frequencies outside the frequency range of the acquisition, the identified mode shapes do not provide any information about the characteristics of the response to those frequencies, which nevertheless contribute to the measured signals;
- they are limited to the degrees of freedom considered and to the identified mode shapes;
- it is also often difficult, in the measurement frequency range, to identify modes that are closely spaced in frequency. This difficulty is greater when

the frequency of those modes is higher because modal density increases with frequency.

In this chapter, the following simplifications are assumed:

• the models have a linear behavior and, therefore, their dynamic response can be described by ordinary differential equations;
• damage is modeled by decreasing the stiffness while keeping the mass constant.

The analysis that follows uses several numerical examples. For that, an FEM of a cantilever beam (Figure 1) with proportional viscous damping, $\mathbf{C} = \beta \cdot \mathbf{K} + \gamma \cdot \mathbf{M}$ with $\beta = 5 \times 10^{-6}$ s and $\gamma = 12$ s^{-1}, is calculated.

The beam is discretized with 20 Timoshenko beam elements, with two nodes each and three degrees of freedom at each node (u_x, u_y, θ_z). The beam dimensions (in mm) are $1800 \times 35 \times 7$ $(L \times b \times h)$, Young's modulus is 185 GPa, the shear modulus is 80 GPa, the shear deflection constant is 6/5, the cross-section area is $A = b \times h = 245$ mm^2 and the mass density is 7917 kg/m^3.

Only the translations along the DOF u_y, that is the direction of the smallest dimension, h, are considered, which means 20 measurement locations, or nodes. The measured quantities are accelerations to simulate the measurement with accelerometers. The forced vibration is obtained using only one excitation force at coordinate 4.

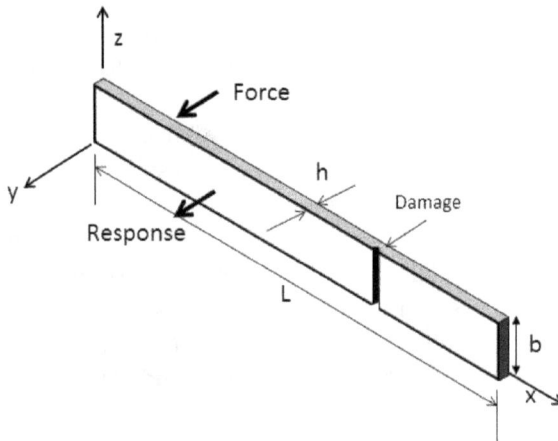

Figure 1.   Beam setup.

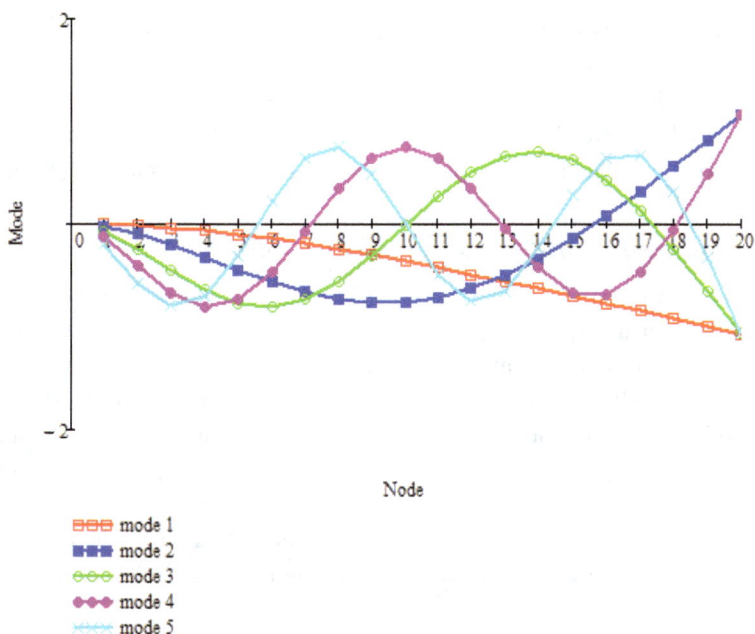

Figure 2.   Cantilever beam. First five mode shapes of the undamaged beam.

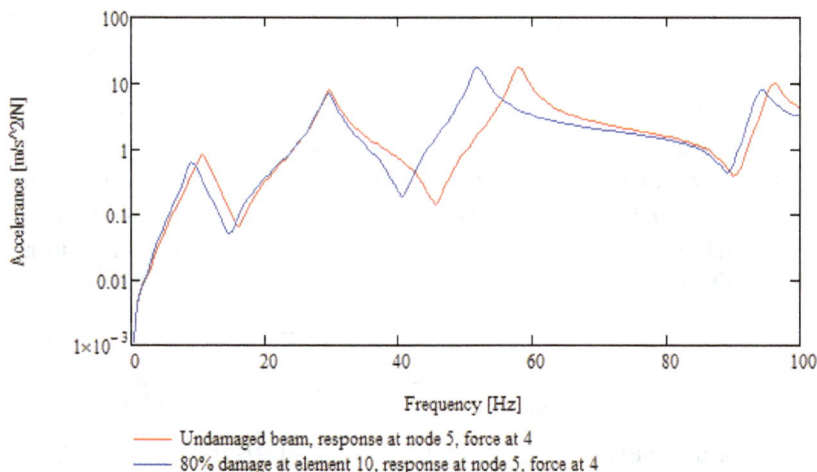

Figure 3.   Cantilever beam with proportional viscous damping. FRF $\alpha_{5,4}(\omega)$ for undamaged and damaged beam.

The damage is simulated with a reduction in the second moment of area of the transverse section at the considered element(s). The frequency analysis is done in the band $[0$–$100]$ Hz with a resolution of $\Delta f = 0.5$ Hz, which means 200 frequency lines, $k = 1 \ldots 200$. The mode shapes of the beam (Figure 2) are normalized to mass matrix. Two FRFs are presented at Figure 3.

2. Comparison Between Mode Shapes Versus ODSs

The importance of obtaining the sensitivity of the mode shapes and ODSs of a system with respect to the changes in its parameters, like when there is damage, lies in the fact that it lets us to assess their ability for damage identification. For this purpose, several numerical simulations are done varying the level of damage, the damage location (element number) and the mode shape or the frequency. For the sake of clarity, in this section, 10 levels of damage have been considered (level 1 is the healthy beam, level 2 is the beam with 10% reduction in the second moment of area of the considered element, and so on until level 10, which is the beam with 90% reduction in the second moment of area.

The mode r of the healthy and damaged beams is compared using the MAC:

$$^d\text{MAC}_r = \frac{\left| \sum_i^d \psi_{i,r} \overline{\psi_{i,r}} \right|^2}{\sum_i^d \psi_{i,r} \overline{^d\psi_{i,r}} \sum_i \psi_{i,r} \overline{\psi_{i,r}}} \tag{11}$$

where $\bar{\psi}$ is the conjugate of the respective function value and d stands for damaged beam.

The ODSs at each frequency ω and force at coordinate $j = 4$, obtained directly from the FRFs, are compared using the frequency domain assurance criterion (FDAC),[9–11] which is nothing more than the generalization of MAC to ODSs

$$^d\text{FDAC}(\omega) = \frac{\left| \sum_i^d \alpha_{i,j}(\omega) \overline{\alpha_{i,j}(\omega)} \right|^2}{\sum_i^d \alpha_{i,j}(\omega) \overline{^d\alpha_{i,j}(\omega)} \sum_i \alpha_{i,j}(\omega) \overline{\alpha_{i,j}(\omega)}} \tag{12}$$

The curvatures of the modes and ODSs can also be compared by MAC and FDAC with the obvious adaptations

$$^d\text{MAC}''_r = \frac{\left| \sum_i^d \psi''_{i,r} \overline{\psi''_{i,r}} \right|^2}{\sum_i^d \psi''_{i,r} \overline{^d\psi''_{i,r}} \sum_i \psi''_{i,r} \overline{\psi''_{i,r}}} \tag{13}$$

and

$$^d\mathrm{FDAC}''(\omega) = \frac{\left| \sum_i^d \alpha_{i,j}''(\omega) \, \overline{\alpha_{i,j}''(\omega)} \right|^2}{\sum_i^d \alpha_{i,j}''(\omega) \, \overline{{}^d\alpha_{i,j}''(\omega)} \sum_i \alpha_{i,j}''(\omega) \, \overline{\alpha_{i,j}''(\omega)}} \tag{14}$$

All the presented indicators vary between $[1\ 0]$, where 1 means no variation and 0 m maximum variation, but in this section the results are presented in a scale of 0–100%, where 0% means no variation and 100% means maximum variation, or, total correlation and no correlation at all, respectively.

2.1. *Comparison of the sensitivity of the mode shapes and that of the mode shapes curvatures to damage*

Figures 4–8 summarize the results of the FEM study for the mode shapes and for the mode shape curvatures (MSCs). From the results, one can conclude the following[a]:

- for the same mode, the mode change grows with the damage level;
- major changes in modes are observed when the damage is located near the antinodes;

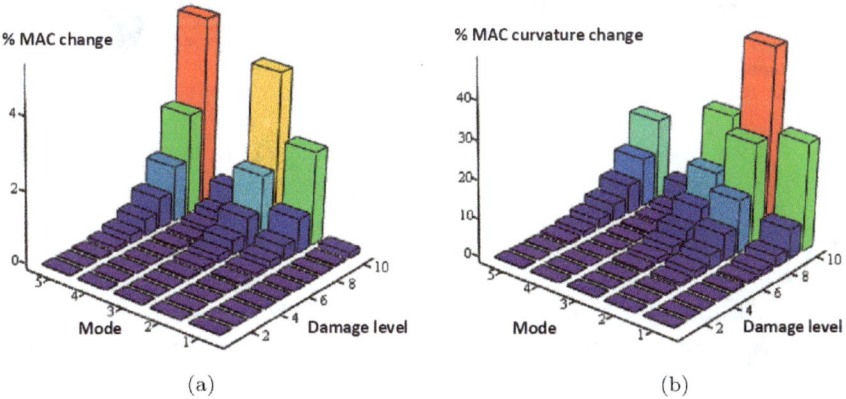

(a)   (b)

Figure 4. (a) Sensitivity of the first five mode shapes and (b) of the first five mode shapes curvatures to 10 levels of damage (damaged element 13).

---

[a]These conclusions are strictly valid for the displayed examples, however it is plausible to admit that this is a general trend.

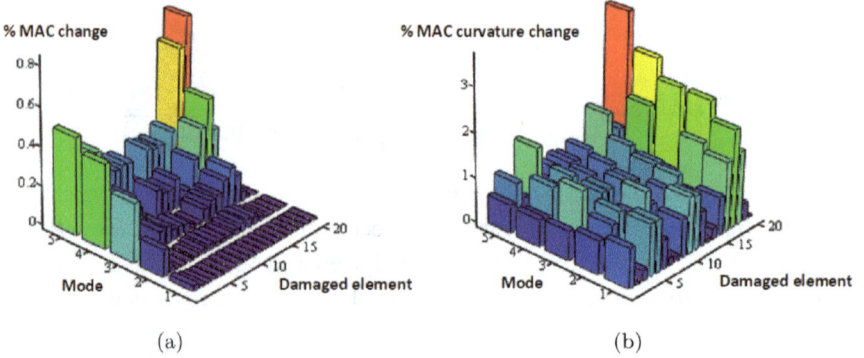

Figure 5. (a) Sensitivity of the first five mode shapes and (b) of the first five mode shapes curvatures to damage localization (50% reduction in the second moment of area at each element).

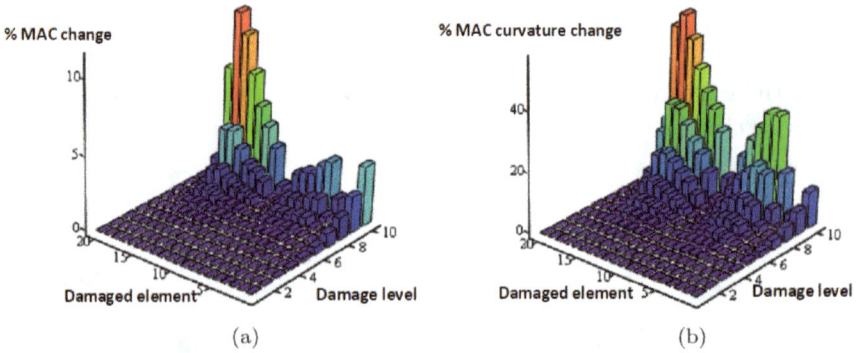

Figure 6. (a) Sensitivity of the third mode shape and (b) of the third mode shape curvature to damage location and damage level.

- changes in modes are practically null when the damage is located near the nodes;
- changes in modes grow with higher modes;
- MSCs, in the presence of damage, behave in the same way as the mode shapes;
- In general, curvatures tend to be more sensitive than mode shapes.

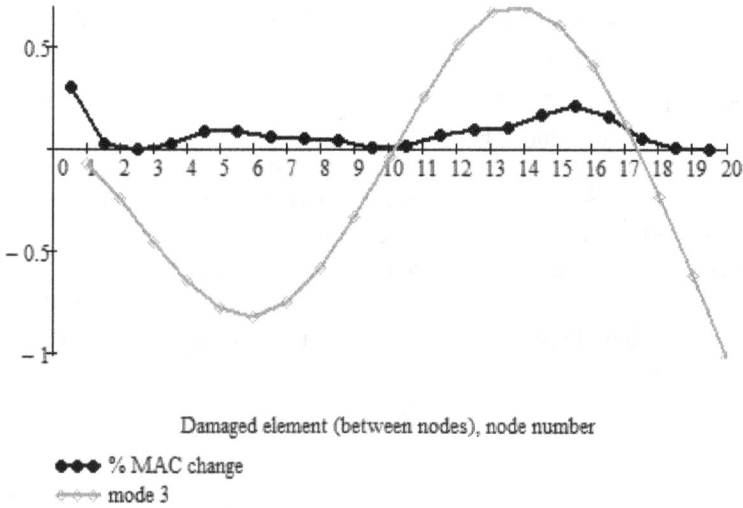

Damaged element (between nodes), node number

●●● % MAC change
◇—◇ mode 3

Figure 7. Variation of the third mode due to 50% damage at element 13.

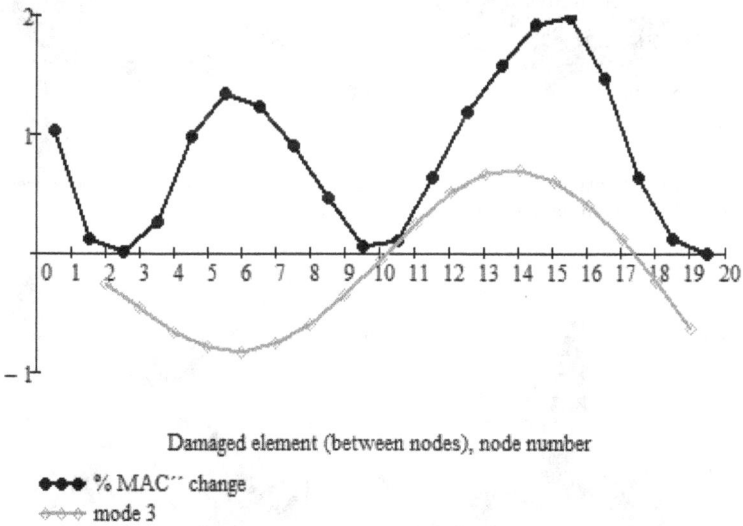

Damaged element (between nodes), node number

●●● % MAC'' change
◇—◇ mode 3

Figure 8. Variation of the third mode curvature to 50% damage at element 13.

2.2. Comparison of the sensitivity of the ODSs and of the ODSs curvatures to damage

Figures 9–13 summarize the results of the FEM study for ODSs. From the results, one can conclude the following:

- changes in ODSs with damage vary greatly with the frequency;
- ODSs tend to be more sensitive to damage than modes;
- ODS curvatures, in the presence of damage, behave the same way as the ODSs;
- it seems that ODS curvatures are more sensitive than the ODSs themselves.

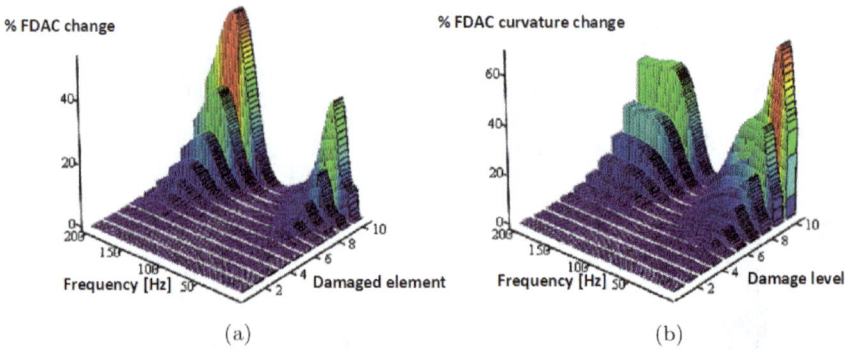

(a)   (b)

Figure 9.   (a) Sensitivity of the ODS and of the ODS curvatures (b), in a frequency band of [0–100] Hz, to damage at element 13.

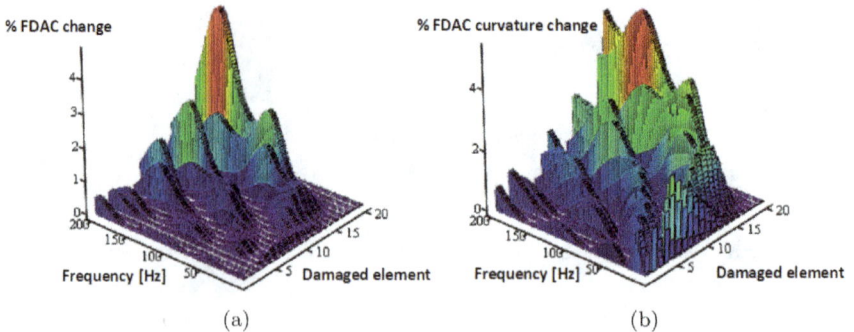

(a)   (b)

Figure 10.   (a) Sensitivity of the ODSs and of the ODSs curvatures (b) to damage localization (50% reduction in the second moment of area at each element).

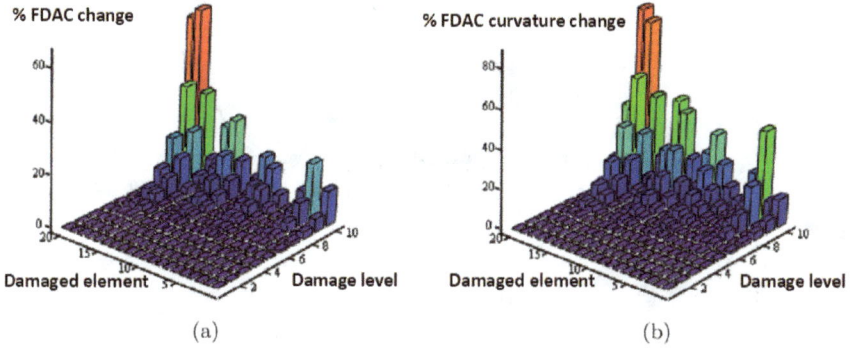

(a)   (b)

Figure 11.   Sensitivity of the 100 Hz ODS (a) and of the 100 Hz ODS curvature (b) to damage location and damage level.

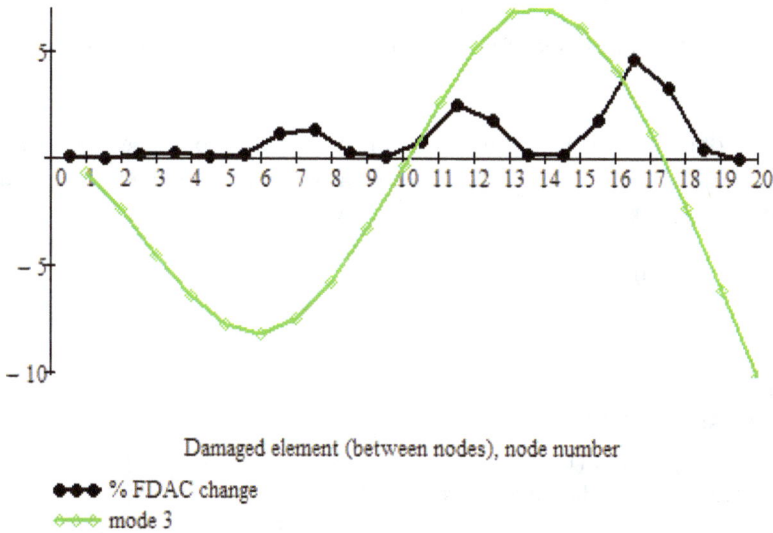

Damaged element (between nodes), node number

●●● % FDAC change
◆◆◆ mode 3

Figure 12.   Variation of the ODS, at frequency 82.5 Hz, to 50% damage at element 13.

3. Damage Identification Methods Based on Response Functions in Time and Frequency Domains

In the following sections, the algorithms of the selected methods are presented. These methods were chosen among the most referenced and also by the heterogeneity of their foundations. Some methods in the

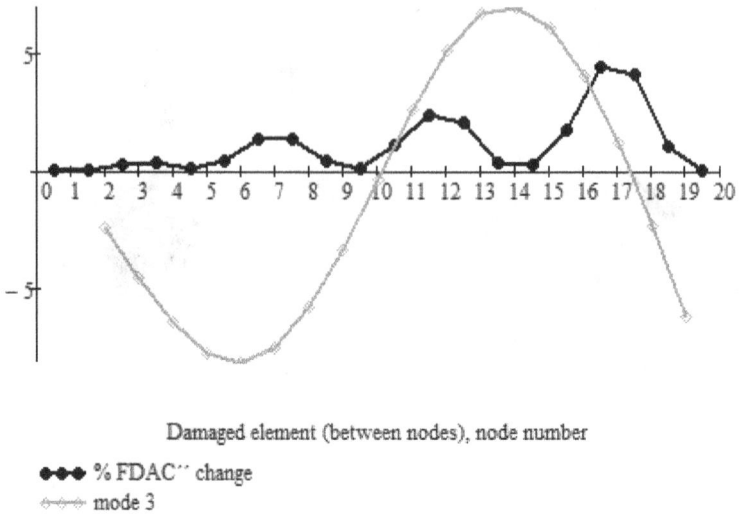

Damaged element (between nodes), node number

●●● % FDAC'' change
◇◇◇ mode 3

Figure 13. Variation of the ODS curvature, at frequency 82.5 Hz, to 50% damage at element 13.

modal domain are also presented for comparison purposes as some of them are the inspiration for those in time and frequency domains. They are:

- Modal domain

  ○ MAC;
  ○ Coordinate MAC;
  ○ Mode shape (MS) method;
  ○ (MSC) method;
  ○ Damage index (DI);
  ○ Strain energy damage index (SEDI).

- Frequency domain

  ○ FRF-based MS method;
  ○ FRF-based MSC method;
  ○ FRF-based DI;
  ○ FRF-based SEDI;
  ○ Resonant gapped smoothing broadband (RGSB);
  ○ FDAC;
  ○ Transmissibility Damage Indicator;
  ○ Error in the constitutive relations.

- Time domain
  - Time domain curvature resonant gapped smoothing method;
  - Time domain fractal dimension index.

3.1. *Modal domain methods*

3.1.1. *Modal assurance criterion*

The MAC, proposed by Allemand,[29] can be applied for damage detection and is defined by the correlation of the mode shapes r of the undamaged and damaged structure

$$^d\text{MAC}_r = \frac{\left|\sum_i^d \psi_{i,r}\overline{\psi_{i,r}}\right|^2}{\sum_i^d \psi_{i,r}\overline{^d\psi_{i,r}} \sum_i \psi_{i,r}\overline{\psi_{i,r}}} \tag{15}$$

3.1.2. *Coordinate modal assurance criterion*

The coordinate MAC, proposed by Lieven,[30] is a method for damage localization and is defined by the correlation

$$^d\text{COMAC}_i = \frac{\left|\sum_r^d \psi_{i,r}\overline{\psi_{i,r}}\right|^2}{\sum_r^d \psi_{i,r}\overline{^d\psi_{i,r}} \sum_r \psi_{i,r}\overline{\psi_{i,r}}} \tag{16}$$

3.1.3. *Mode shape method*

The MS method, proposed by Ho and Ewins,[12] is based on the supposition that the major absolute differences, between the mode shape of the damaged structure and the mode shape of the healthy structure, appear at the nodes, or measurement locations, closer to the damaged region. Originally defined for localization, it can be used for detection as well. The index is the absolute difference calculated at each identified mode r

$$^d\text{MS}_{i,r} = \left|^d\psi_{i,r} - \psi_{i,r}\right| \tag{17}$$

3.1.4. *Mode shape curvature method*

The MSC method, first developed by Pandey, Biswas and Samman[13] and later improved[b] by Wahab and Roeck,[27] is based on the same assumption

---

[b]Known as Curvature Damage Factor.

of the MS method, although using the curvatures of the mode shapes. The location of the damage is assessed by the largest computed absolute difference between the ASCs of the damaged and undamaged structure, as follows:

$$^d\text{MSC}_{i,r} = \left| {}^d\psi''_{i,r} - \psi''_{i,r} \right| \tag{18}$$

3.1.5. Damage index

The DI, developed by Stubbs and Kim,[14] is based on the strain energy of the flection modes and also makes use of the curvature changes. Originally defined for localization, it can also be used for detection. For the purpose of this chapter, it was decided to use the following formulation of the DI method[15]:

$$^d\text{DI}_{i,r} = \frac{\left({}^d\psi''^{\,2}_{i,r} + \sum_i {}^d\psi''^{\,2}_{i,r}\right)\sum_i \psi''^{\,2}_{i,r}}{\left(\psi''^{\,2}_{i,r} + \sum_i \psi''^{\,2}_{i,r}\right)\sum_i^d \psi''^{\,2}_{i,r}} \tag{19}$$

3.1.6. Strain energy damage index

The SEDI, developed by Petro et al.[16] is also, as DI, based on the strain energy. Once more, it was originally defined for localization but can be used for detection too. The index is

$$^d\text{SEDI}_{i,r} = \frac{\left| {}^d\psi''^{\,2}_{i,r} + {}^d\psi''^{\,2}_{i+1,r} - \psi''^{\,2}_{i,r} - \psi''^{\,2}_{i+1,r} \right|}{\left| \psi''^{\,2}_{i,r} + \psi''^{\,2}_{i+1,r} \right|} \tag{20}$$

3.2. Frequency domain methods

Based upon the same assumption made for most of the modal domain methods mentioned earlier, i.e. the damage is located at the point where the change of the index is the highest for a mode shape or group of mode shapes, the authors also suggested[15,19] that the same assumption could be generalized to the entire frequency range, i.e. the damage is located at the point where the change in the index is the highest for FRF-based ODS. Subsequently, these methods are identified as *FRF_ name of original mode shape method*.

When calculating the damage indices for the FRF-based methods, as we go along the frequency range, adding more and more information, the results begin to degenerate instead of improving.[20] The reason for this is that close to the resonances and anti-resonances the differences between the damaged and undamaged FRFs, specially for lightly damped structures and/or when damage is big, tend to be high, and if by any chance the method gives a false damage location at such frequencies, the simple summation of such a result to the others at different frequencies may completely mask the true location. To overcome this, at each frequency, we look up for the location where the difference between damage and undamaged cases is a maximum and at that location we count an occurrence; as we proceed along the frequency range, we simply keep up summing the occurrences, not the differences themselves.

This procedure of looking at the maxima and summing up the number of times that they occur at each location can be named as procedure of maximum occurrences (MO). This procedure, for some of the methods, does not count just the maximum but also two or four maxima, depending on the method. This procedure can be successfully applied to both detection and localization methods.

3.2.1. *FRF-based MS method*

The FRF-based mode shape (FRF-MS) method, proposed by Maia *et al.*,[19] is based on the same assumption of the MS method, though generalized to all frequencies of the measured FRFs

$$^{d}\text{FRF\_MS}_i(\omega) = \left| {}^{d}\alpha_{i,j}(\omega) - \alpha_{i,j}(\omega) \right| \tag{21}$$

First developed as a localization method, it was recently proposed as a detection method,[23] with the name of frequency domain ODS difference indicator.

3.2.2. *FRF-based mode shape curvature method*

This method, proposed by Sampaio *et al.*,[15] is the generalization of the MSC method to the FRF-based ODSs; it can be used for detection and localization, and is defined by

$$^{d}\text{FRF\_MSC}_i(\omega) = \left| {}^{d}\alpha''_{i,j}(\omega) - \alpha''_{i,j}(\omega) \right| \tag{22}$$

3.2.3. *FRF-based damage index*

This method, proposed by Maia *et al.*,[19] is the generalization of the DI method to the FRF-based ODSs and it can be used for both detection and localization

$$^d\text{FRF\_DI}_i(\omega) = \left| \frac{\left({}^d\alpha''^{\,2}_{i,j}(\omega) + \sum_i {}^d\alpha''^{\,2}_{i,j}(\omega) \right) \sum_i \alpha''^{\,2}_{i,j}(\omega)}{\left(\alpha''^{\,2}_{i,j}(\omega) + \sum_i \alpha''^{\,2}_{i,j}(\omega) \right) \sum_i {}^d\alpha''^{\,2}_{i,j}(\omega)} \right| \tag{23}$$

3.2.4. FRF-based strain energy damage index

The generalization of the SEDI method to the FRF-based ODSs can be also used for detection and localization, and is defined by

$$^d\text{FRF\_SED}_i(\omega)$$
$$= \left| \frac{{}^d\alpha''^{\,2}_{i,j}(\omega) + {}^d\alpha''^{\,2}_{i+1,j}(\omega) - \alpha''^{\,2}_{i,j}(\omega) - \alpha''^{\,2}_{i+1,j}(\omega)}{\alpha''^{\,2}_{i,j}(\omega) + \alpha''^{\,2}_{i+1,j}(\omega)} \right| \tag{24}$$

3.2.5. Resonant gapped-smoothing broadband method

The RGSB method developed by Ratcliffe[17,18,24] uses the curvatures of the FRF-based ODSs to localize damage. This method does not need a baseline, a measurement or a numerical model of the undamaged structure, to compare with the measurement of the damaged structure. RGSB searches for discontinuities in the frequency domain ODS curvatures in the damaged structure. The technique for highlighting the lack of smoothness, and hence the location of damage, is the evaluation of the absolute difference between the values of third-order polynomials

$$^\omega P_3(i) = a_0 + a_1 i + a_2 i^2 + a_3 i^3 \tag{25}$$

where the coefficients a_0, a_1, a_2, a_3 are calculated for each set of four curvature values of the ODS at the measurement points:

- (3, 4, 5, 6) if $i = 2^c$;
- (2, 4, 5, 6) if $i = 3$;
- $(N - 5, N - 4, N - 3, N - 2)$ if $i = N - 1$;
- $(N - 5, N - 4, N - 3, N - 1)$ if $i = N - 2$;
- $(i - 2, i - 1, i + 1, i + 2)$ for all the other values of i.

and the curvatures values at the measurement point i.

---

[c]Because the second-order central difference equation uses $i-1$, i and $i+1$ points to calculate the curvature, location 2 is the first location to have the curvature evaluated.

Since the FRF values are usually complex, the squares of these differences are calculated for the real and for the imaginary parts and added together as follows:

$$^d\mathrm{RGSB}_i(\omega) = \left[\mathrm{Re}(^\omega P_3(i)) - \mathrm{Re}\left(\alpha''_{i,j}(\omega)\right)\right]^2$$
$$+ \left[\mathrm{Im}(^\omega P_3(i)) - \mathrm{Im}\left(\alpha''_{i,j}(\omega)\right)\right]^2 \tag{26}$$

3.2.6. *Frequency domain assurance criterion*

The FDAC, developed by Pascual[9,10] (and later by Heylen *et al.*,[28] with the response vector assurance criterion, and by Sampaio *et al.*,[21] with the detection and relative quantification indicator (DRQ), can be used as a detection index and is defined by

$$^d\mathrm{FDAC}(\omega) = \frac{\left|\sum_i^d \alpha_{i,j}(\omega)\overline{\alpha_{i,j}(\omega)}\right|^2}{\sum_i^d \alpha_{i,j}(\omega)\overline{^d\alpha_{i,j}(\omega)}\sum_i \alpha_{i,j}(\omega)\overline{\alpha_{i,j}(\omega)}} \tag{27}$$

3.2.7. *Transmissibility damage indicator*

The transmissibility damage indicator (TDI), developed by Almeida *et al.*,[22] is a detection index and is defined by

$$^d\mathrm{TDI}(\omega) = \frac{\left|\sum_i \frac{^d\alpha_{i,j}(\omega)}{^d\alpha_{i+1,j}(\omega)}\overline{\frac{\alpha_{i,j}(\omega)}{\alpha_{i+1,j}(\omega)}}\right|^2}{\left(\sum_i \frac{^d\alpha_{i,j}(\omega)}{^d\alpha_{i+1,j}(\omega)}\overline{\frac{^d\alpha_{i,j}(\omega)}{^d\alpha_{i+1,j}(\omega)}}\right)\left(\sum_i \frac{\alpha_{i,j}(\omega)}{\alpha_{i+1,j}(\omega)}\overline{\frac{\alpha_{i,j}(\omega)}{\alpha_{i+1,j}(\omega)}}\right)} \tag{28}$$

3.2.8. *Error in the constitutive relations*

A different damage localization method can be cast upon the definition of a function which gives a measure of the error in the constitutive relations (ECR). This measure was proposed by Ladevèze[31] assuming that both boundary and equilibrium equations can be considered as "reliable" equations, whereas the constitutive relations are less "reliable", as they request the knowledge of the material properties of the structure. In this context, one may assume two constitutive relations: the first relation is given by

$$\sigma = \mathbf{E}\,\varepsilon(\mathbf{u}) + \mathbf{B}\dot{\varepsilon}(\mathbf{u}) \tag{29}$$

where σ is the stress tensor, with one term proportional to the strain tensor ε through the Hooke's operator $\mathbf{E}$ and a dissipative term proportional to the strain velocity $\dot{\varepsilon}$, through an operator $\mathbf{B}$, related to viscous damping. The second constitutive relation is given by

$$\mathbf{\Gamma} = \rho\ddot{\mathbf{u}} + \mathbf{A}\dot{\mathbf{u}} \tag{30}$$

where $\mathbf{\Gamma}$ is the inertia force density, with one term proportional to the mass density ρ and a term related to dissipative forces through an operator $\mathbf{A}$.

In general, one has a theoretical model which does not comply exactly with the experimental results; as a consequence, if there are errors in the theoretical model, it makes sense to focus our attention on the constitutive relations. Regarding the errors and lack of knowledge in the modeling parameters, if one considers an experimental displacement field, the constitutive relations are not exactly attained due to the fact that those constitutive relations depend on "static" quantities, the material properties, and on "kinematic" ones, the displacement $\mathbf{u}$ ($\mathbf{u} = \mathbf{u}_k$). Therefore, one can introduce "kinematic" constitutive relations that are attained regarding the dynamic displacements, as well as "static" displacement fields, that allow the verification of the defined constitutive relations. Hence, one must find the displacement fields that verify a set of "kinematic" constitutive relations

$$\sigma_k = \mathbf{E}\,\varepsilon(\mathbf{u}_k) + \mathbf{B}\dot{\varepsilon}(\mathbf{u}_k) \tag{31}$$

$$\Gamma_k = \rho\ddot{\mathbf{u}}_k + \mathbf{A}\dot{\mathbf{u}}_k \tag{32}$$

and a set of "static" ones

$$\sigma_s = \mathbf{E}\,\varepsilon(\mathbf{u}_\sigma) + \mathbf{B}\dot{\varepsilon}(\mathbf{u}_\sigma) \tag{33}$$

$$\Gamma_S = \rho\ddot{\mathbf{u}}_\Gamma + \mathbf{A}\dot{\mathbf{u}}_\Gamma \tag{34}$$

where $\mathbf{u}_\sigma$ is the displacement field associated to the "static" stress field and $\mathbf{u}_\Gamma$ is the displacement field associated to the "static" inertia force density field.

If one considers harmonic data in a given frequency range, it is possible to find the space of admissible displacement fields that minimizes the error in the constitutive relations. Additionally, if one takes into account experimental data, one must modify the error in the constitutive relations by adding an extra term, which quantifies the difference between the experimental data, denoted by $\tilde{\bullet}$, and their corresponding theoretical counterparts.

Using a numerical discretization and for a simplified version, neglecting damping effects and considering only errors related to elastic forces, the

ECR can be given by

$$
\begin{aligned}
E_\omega'^2 =\ & (\boldsymbol{\alpha}_U - \boldsymbol{\alpha}_V)^T \mathbf{K} (\boldsymbol{\alpha}_U - \boldsymbol{\alpha}_V) \\
& + \frac{r}{1-r} \left[(\Pi\boldsymbol{\alpha}_U - \tilde{\boldsymbol{\alpha}})^T \mathbf{K_R} (\Pi\boldsymbol{\alpha}_U - \tilde{\boldsymbol{\alpha}}) \right]
\end{aligned}
\tag{35}
$$

where $\boldsymbol{\alpha}_U$ and $\boldsymbol{\alpha}_V$ are FRFs obtained for the admissible displacement fields that satisfy the static and kinematic constitutive relations related to the stress fields; $r \in [0, 1]$ is a weighting parameter which indicates the quality of the experimental data (lower values for noisy data); Π is a projection operator mapping the numerical quantities onto the measured coordinates and the subscript R defines reduced or condensed quantities.

The FRFs $\boldsymbol{\alpha}_U$ and $\boldsymbol{\alpha}_V$ are computed by the minimization of the defined ECR, such that

$$
\frac{\partial E_\omega'^2}{\partial S_{\text{adm}_i}} = 0
\tag{36}
$$

with $S_{\text{adm}} = [\boldsymbol{\alpha}_U, \boldsymbol{\alpha}_V]$ subject to the dynamic equilibrium equation $\omega^2 \mathbf{M}\boldsymbol{\alpha}_V = [\mathbf{K} - \mathbf{Z}]\boldsymbol{\alpha}_U$, where $\mathbf{Z}$ is the dynamic stiffness matrix. In such a case, the admissible solution set that minimizes the ECR is given by

$$
\boldsymbol{\alpha}_U = \left[\mathbf{Z} \left[\mathbf{K}^{-1}\mathbf{Z} + \Pi^T\mathbf{K_R}\Pi \right] \right]^{-1} \left[\mathbf{Z}\mathbf{K}^{-1}\underline{\mathbf{I}} + \Pi^T\mathbf{K_R}\Pi\tilde{\boldsymbol{\alpha}} \right]
\tag{37}
$$

$$
\boldsymbol{\alpha}_V = \mathbf{K}^{-1} \left[\underline{\mathbf{I}} + \omega^2 \mathbf{M}\boldsymbol{\alpha}_U \right]
\tag{38}
$$

where $\underline{\mathbf{I}}$ is a vector of zeros except at the coordinate where the force is applied.

The formulation of a damage indicator based on a more generic formulation of the ECR function of the FRFs is detailed in Ref. 32. In that work, one can find the expression for the ECR function of the FRFs, considering viscous damping effects and the possibility of errors in the inertial constitutive relations, such that

$$
\begin{aligned}
E_\omega'^2 =\ & \frac{1-\gamma}{2} (\boldsymbol{\alpha}_U - \boldsymbol{\alpha}_V)^H [\mathbf{K} + 2\pi\omega\mathbf{B}] (\boldsymbol{\alpha}_U - \boldsymbol{\alpha}_V) \\
& + \frac{\gamma}{2} (\boldsymbol{\alpha}_U - \boldsymbol{\alpha}_W)^H \mathbf{M} (\boldsymbol{\alpha}_U - \boldsymbol{\alpha}_W) \\
& + \frac{r}{1-r} \left\{ \frac{1-\gamma}{2} (\Pi\boldsymbol{\alpha}_U - \tilde{\boldsymbol{\alpha}})^H [\mathbf{K_R} + 2\pi\omega\mathbf{B_R}] (\Pi\boldsymbol{\alpha}_U - \tilde{\boldsymbol{\alpha}}) \right. \\
& \left. + \frac{\gamma}{2} (\Pi\boldsymbol{\alpha}_W - \tilde{\boldsymbol{\alpha}})^H \mathbf{M_R} (\Pi\boldsymbol{\alpha}_W - \tilde{\boldsymbol{\alpha}}) \right.
\end{aligned}
$$

$$+\frac{1}{2}\left([\mathbf{K}+i\omega\mathbf{B}]\,\boldsymbol{\alpha}_V-\omega^2\mathbf{M}\boldsymbol{\alpha}_W-\underline{\mathbf{I}}\right)^H$$

$$\times\,\mathbf{K}^{-1}\left([\mathbf{K}+i\omega\mathbf{B}]\,\boldsymbol{\alpha}_V-\omega^2\mathbf{M}\boldsymbol{\alpha}_W-\underline{\mathbf{I}}\right)\Bigg\} \qquad (39)$$

where $\boldsymbol{\alpha}_W$ are the FRFs obtained for the admissible displacement field that satisfies the static constitutive relation related to the inertia force density field, $\gamma \in [0,1]$ is a weighting parameter related to the degree of confidence on the information about the elastic properties, being 0 if the mass matrix is completely known, and the superscript H is the complex conjugated transpose operator.

Note that the admissible solution set $S_{\text{adm}} = [\boldsymbol{\alpha}_U, \boldsymbol{\alpha}_V, \boldsymbol{\alpha}_W]$ is obtained by the minimization of the ECR. With the solution to this minimization problem, one can define a relative error at each frequency for each finite element of the model and for each applied force k. Denoting the FRFs at the degrees of freedom of each element or group of elements j as $\boldsymbol{\alpha}_{\bullet_{jk}}$ and considering proportional damping, the local relative error at each frequency may be given by

$$E'^2_{jk\omega} = \left[(1-\gamma)\,(1+2\pi\omega\beta)\left(\boldsymbol{\alpha}_{U_{jk}}-\boldsymbol{\alpha}_{V_{jk}}\right)^H \mathbf{K}_j\left(\boldsymbol{\alpha}_{U_{jk}}-\boldsymbol{\alpha}_{V_{jk}}\right)\right.$$

$$\left.+\gamma\omega^2\left(\boldsymbol{\alpha}_{U_{jk}}-\boldsymbol{\alpha}_{W_{jk}}\right)^H \mathbf{M}_j\left(\boldsymbol{\alpha}_{U_{jk}}-\boldsymbol{\alpha}_{W_{jk}}\right)\right]$$

$$\times\left[\frac{(1-\gamma)}{2}\,(1+2\pi\omega\beta)\left(\boldsymbol{\alpha}_{U_k}^H\mathbf{K}\boldsymbol{\alpha}_{U_k}+\boldsymbol{\alpha}_{V_k}^H\mathbf{K}\boldsymbol{\alpha}_{V_k}\right)\right.$$

$$\left.+\frac{\gamma}{2}\omega^2\left(\boldsymbol{\alpha}_{U_k}^H\mathbf{M}\boldsymbol{\alpha}_{U_k}+\boldsymbol{\alpha}_{W_k}^H\mathbf{M}\boldsymbol{\alpha}_{W_k}\right)\right]^{-1} \qquad (40)$$

Note that the numerator of $E'^2_{jk\omega}$ reflects the error, at the element level, in the viscoelastic and kinetic energies weighted by the parameter γ, whereas the denominator represents the viscoelastic and kinetic energies of the entire structure. Thus, the error in the viscoelastic and kinetic energies is related to the extent of damage at each element and it is convenient to normalize it so that the error stays between 0 and 1.

3.3. Time domain methods

3.3.1. Time domain resonant gapped smoothing method

The time domain resonant gapped smoothing (TRGS) method is a new method for damage detection and damage localization. It is based on the RGSB method but applied to time ODS curvatures. The interpolating polynomial can be of degree one, two, or three. This method has several advantages:

- it is a real-time index;
- it only needs the time responses;
- it does not need frequency analysis;
- it does not need modal identification;
- it does not need baseline, a measurement or a model, as a reference.

3.3.2. *Time domain Katz fractal dimension method*

The time domain Katz fractal dimension (TKFD) method is another new method for damage detection. Basically, the fractal dimension is calculated for each time ODS with the algorithm proposed by Katz.[26] Adding all the fractal numbers calculated for each time ODS, obtained for an acquisition, one ends up with an index. Because the fractal dimension is a measure of the smoothness of a waveform,[25] it is expected that this index grows with the damage whose effect is to turn the time ODS sharper at its location.

The advantages of this method are as follows:

- easy implementation;
- light computation;
- real-time index;
- it only needs the time responses;
- it does not need frequency analysis;
- it does not need modal identification;
- it does not need baseline, a measurement or a model, as a reference.

4. Detection and Localization Capabilities of the Presented Indicators

The following sections present the results of the indicators, introduced in Section 3, for the following numerical simulations:

- Damage detection — calculation of the detection indicators for each situation of one damage in one beam element considering five damage levels, i.e. 1 — no damage, 2 — 20 %, 3 — 40%, 4 — 60% and 5 — 80% reduction in the second moment of area and all possible locations (beam elements);
- Damage localization — calculation of the localization indicators at each node for each situation of one damage in one beam element considering

just one damage level, i.e. 50% reduction in the second moment of area, and all possible locations.

To facilitate the comparison between different methods, they are redefined without losing the identity of their original definitions for detection methods:

(1) for each mode shape, or ODS, and for each node, the index is normalized with respect to the biggest one among the five damage cases (the five indicators are divided by the biggest of them);

(2) then, for each damage normalized case, the average of the index at all locations and at all mode shapes, or at all ODSs, is calculated;

(3) finally, the five results (five damages cases) are adjusted to a scale of [0–100], where 0 means structure with smaller or no damage and 100 means structure with biggest damage. This adjustment is made to use all the values in the scale 0–100, i.e. the smallest value takes 0 and the highest 100. All the other values are changed accordingly;

(4) the results of the detection methods are displayed in a 3D bar plot where the x-axis is the damage level, the y-axis is the damaged element and the z-axis is the index of the method in percentage. If one looks to the plane $y = e$, one can see how the index varies with damage at the element e of the cantilever beam. If one looks to the plane $x = d$, one can see how the index varies with the localization of the damage d. What is expected for a good detection capability is, for each damaged element, the growing amplitude of the bars with the growing damage level. This means that the index recognizes, for any possible localization of the damage, the five levels of damage.

and for localization methods:

(1) for each mode shape, or ODS, and for each damage case, the index is normalized with respect to the biggest one among the 20 measurement locations (the 20 indicators are divided by the biggest of them);

(2) then, for each normalized location (or node) and damage case, the average of the index at all mode shapes, or at all ODSs, is calculated;

(3) finally, the 20 results (20 possible locations) are adjusted to a scale of 0–100, where 0 means location without damage and 100 means location with damage. This adjustment is made to use the entire scale 0–100,

i.e. the smallest value takes 0 and the highest 100. All the other values are changed accordingly;

(4) the results of the localization methods, for the 50% damage case, are displayed in a 2D contour plot where the x-axis is the node (or location) and the y-axis is the damaged element. The index of the method is presented in percentage and with rainbow type colored pixels (dark blue is 0 and red is 100). A method that localizes well will display a stairway diagonal (the 2 nodes of the damaged element should display red and the others dark blue).

4.1. *Modal domain methods*

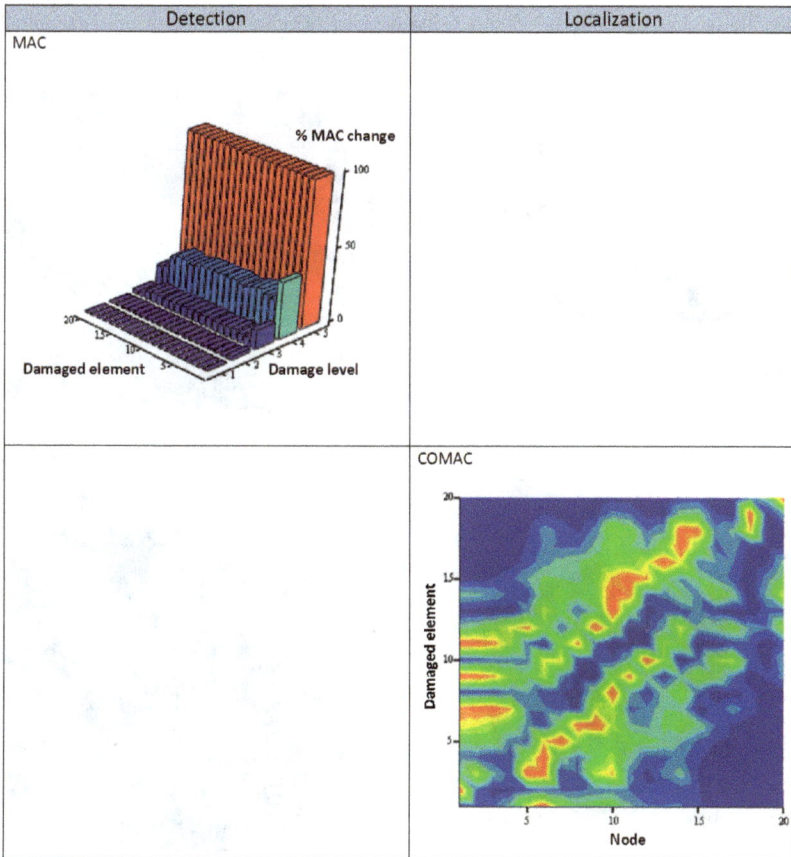

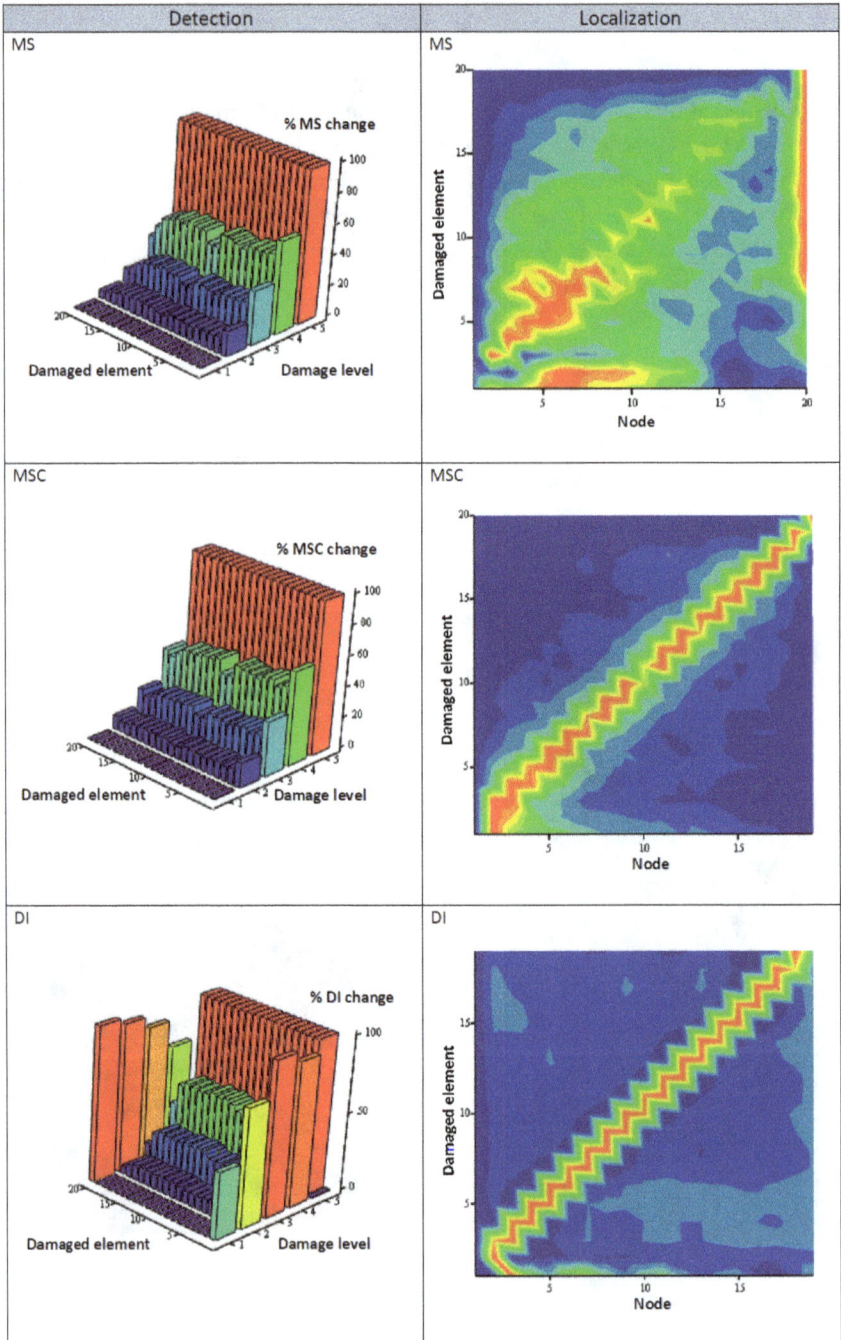

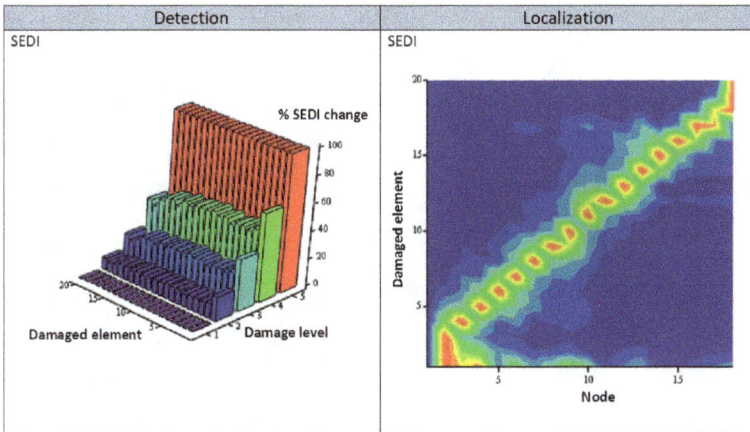

4.2. *Frequency domain methods*

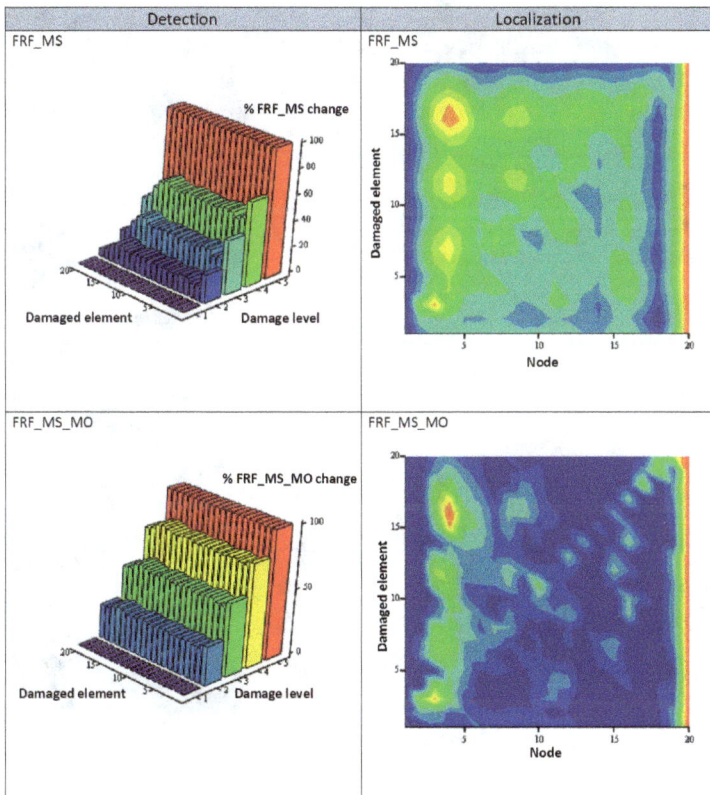

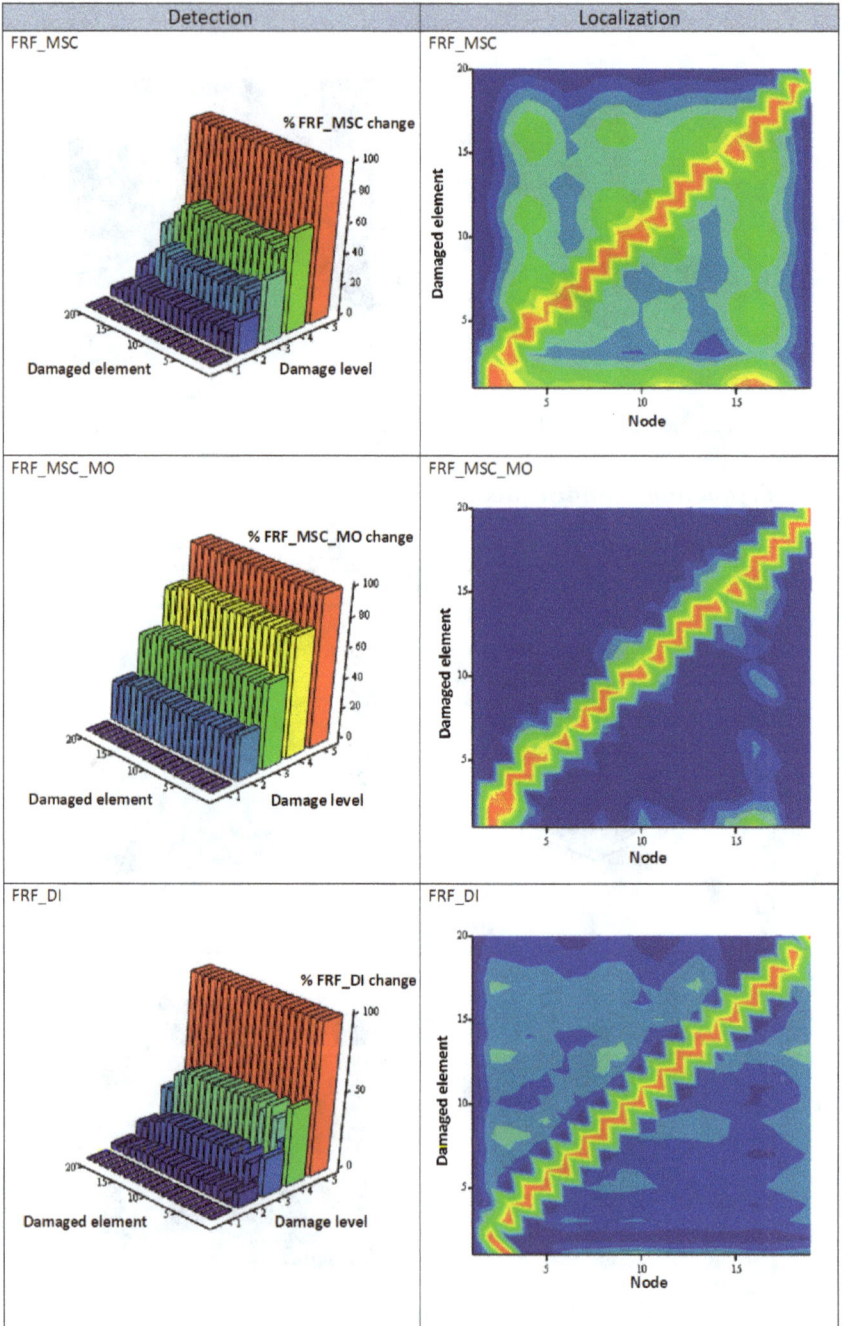

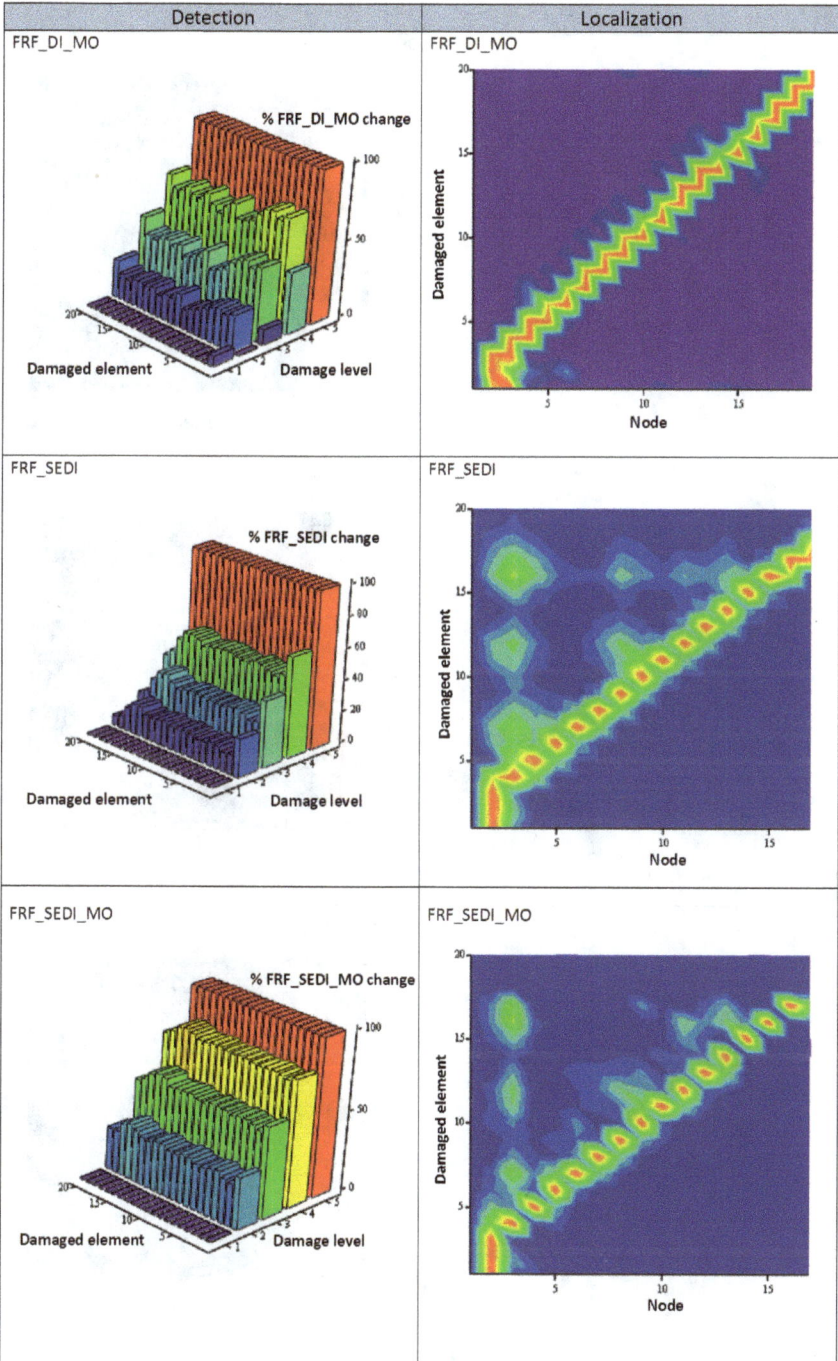

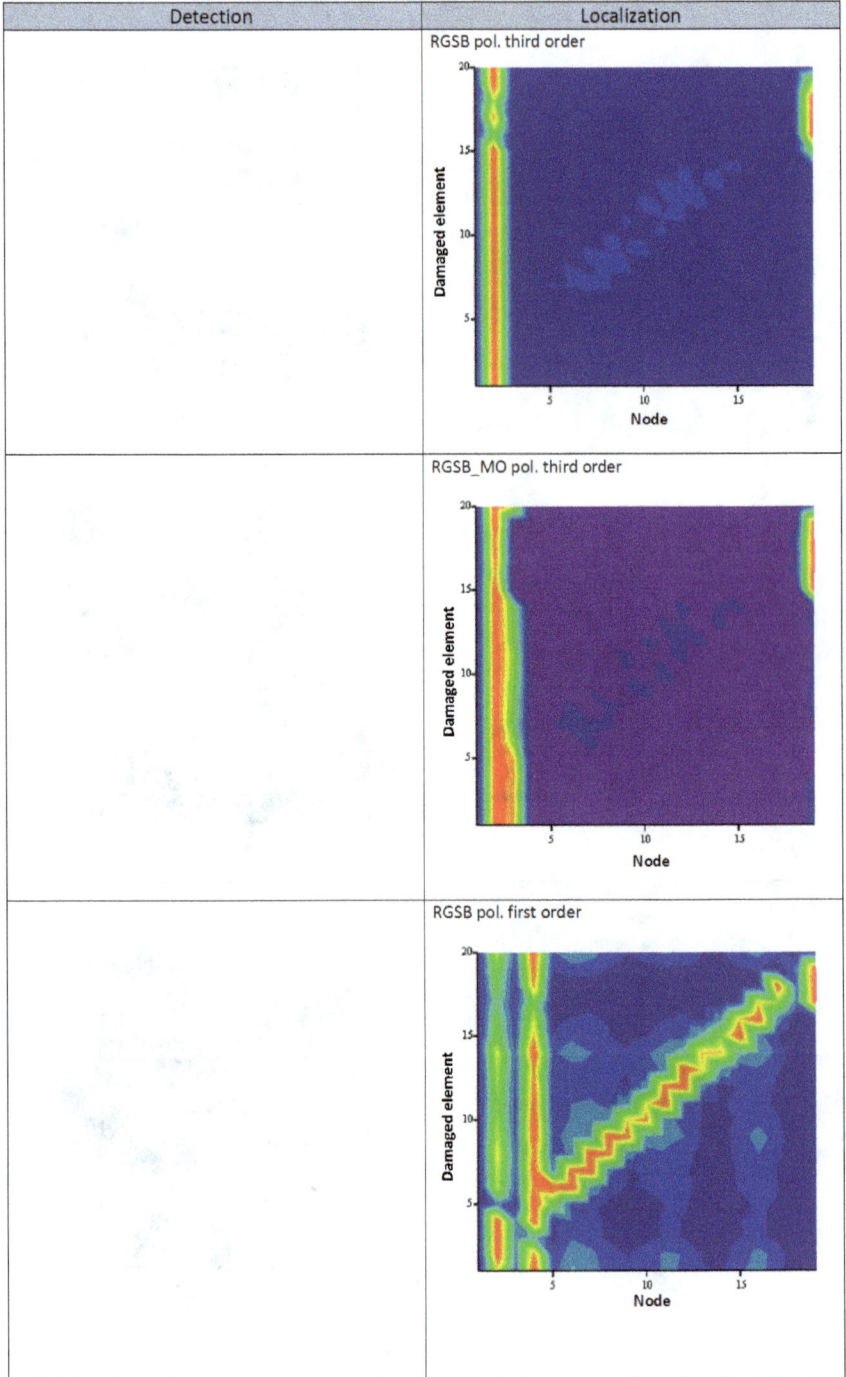

Detection	Localization
	RGSB pol. third order
	RGSB_MO pol. third order
	RGSB pol. first order

Detection	Localization
	RGSB_MO pol. first order
FDAC/RVAC/DRQ	
FDAC/RVAC/DRQ_MO	

Detection	Localization
TDI	

| TDI_MO | |

| ECR | ECR |

4.3. *Time domain methods*

Detection	Localization
TRGS pol. third order, free response	TRGS pol. third order, free response

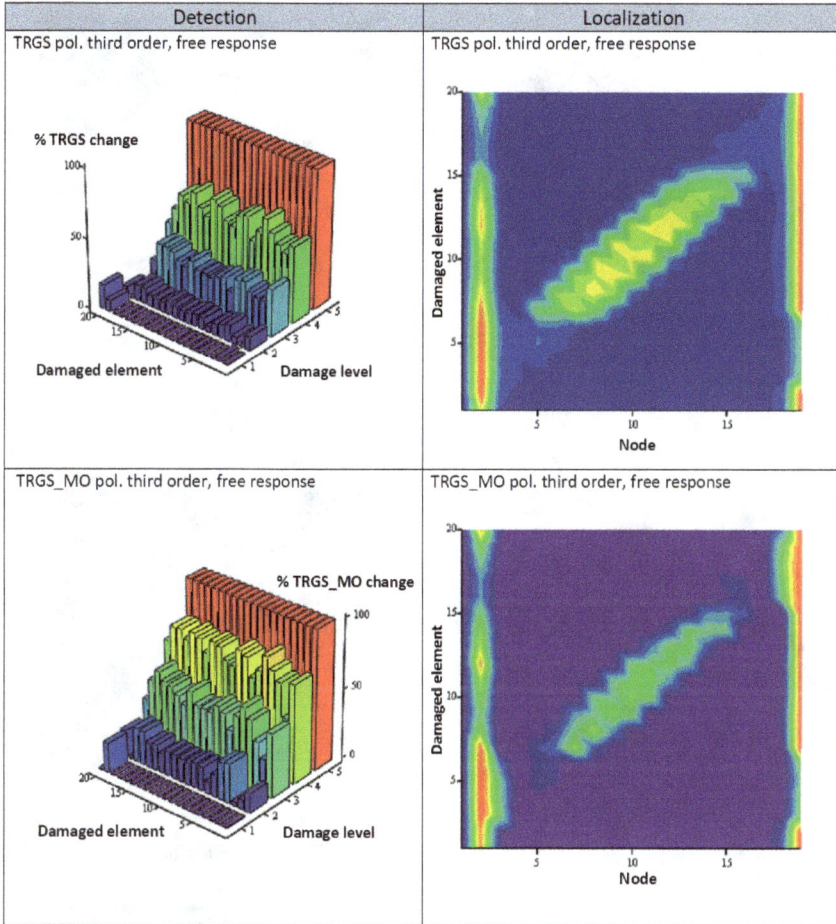

| TRGS_MO pol. third order, free response | TRGS_MO pol. third order, free response |

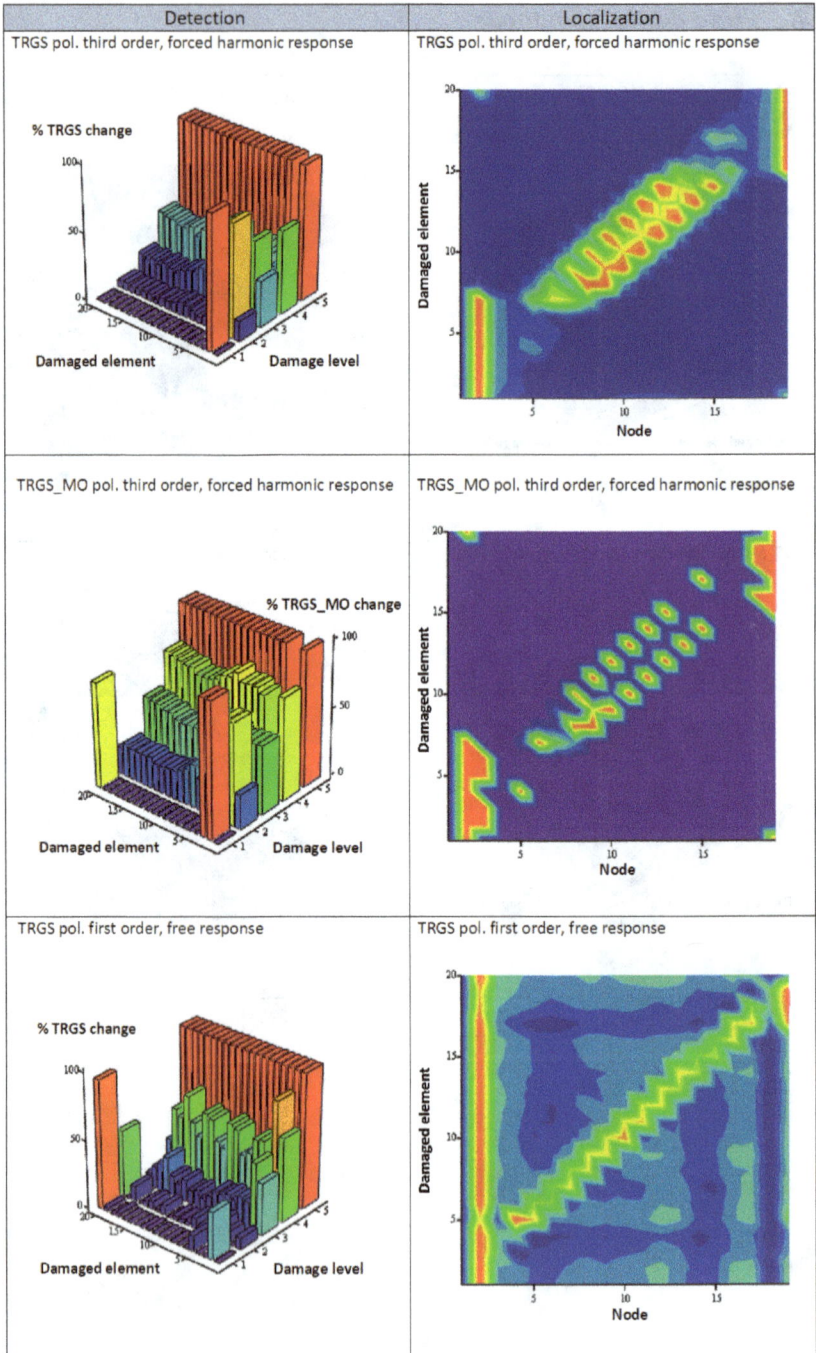

Detection	Localization
TRGS pol. third order, forced harmonic response	TRGS pol. third order, forced harmonic response
TRGS_MO pol. third order, forced harmonic response	TRGS_MO pol. third order, forced harmonic response
TRGS pol. first order, free response	TRGS pol. first order, free response

Detection	Localization
TRGS_MO pol. first order, free response	TRGS_MO pol. first order, free response
TRGS pol. first order, forced harmonic response	TRGS pol. first order, forced harmonic response
TRGS_MO pol. first order, forced harmonic response	TRGS_MO pol. first order, forced harmonic response

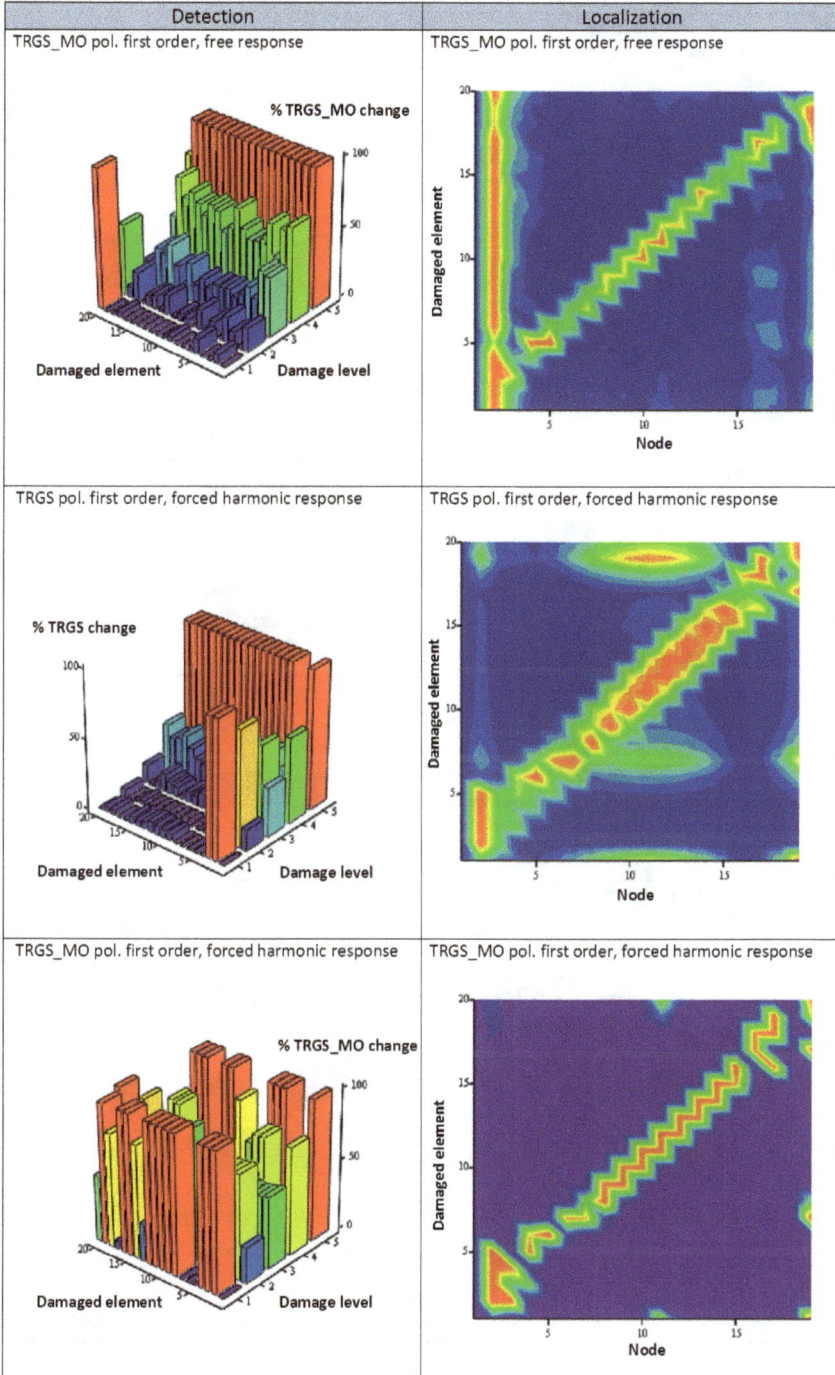

Detection	Localization
TKFD, free response	
TKFD_MO, free response	
TKFD, forced harmonic response	

Detection	Localization
TKFD_MO, forced harmonic response 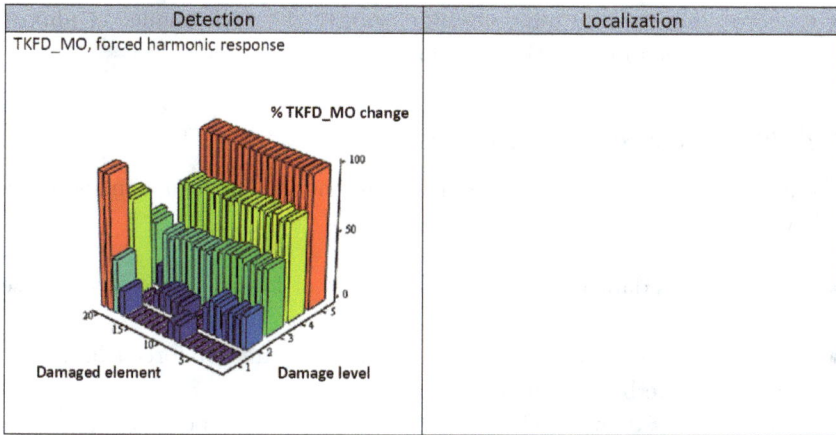	

5. Conclusions

Two sets of numerical tests were executed to evaluate the performance of the presented methods of Section 3, concerning their ability for damage detection and damage localization. Each test simulates a cantilever beam with proportional viscous damping.

In the following sections, conclusions are presented.

5.1. *Modal domain methods*

From the results of the modal domain methods, one can conclude the following:

- all the methods detect damage;
- the index DI did not detect damage at the free end of the beam;
- MSC, DI, and SEDI localize damage except at the fixed end of the beam;
- MS did not localize the damage precisely;
- COMAC did not localize at all.

5.2. *Frequency domain methods*

From the results of the frequency domain methods, one can conclude the following:

- all the methods detect damage;
- FRF_MSC, FRF_DI, FRF_SEDI and ECR localize the damage well;
- RGSB and RGSB MO localize damage with some difficulty;
- RGSB methods localize better when a first-order polynomial is used;

- the procedure of MO improves significantly the performance of almost all the methods in detection and localization.

5.3. *Time domain methods*

From the results of the time domain methods, one can conclude the following:

- TRGS detects damage when it uses the free or forced harmonic response of the beam;
- TRGS localizes damage reasonably well when free or forced harmonic response of the beam is used;
- TKFD detects damage when the forced harmonic response of the beam is used;
- TKFD detects damage badly when the free response of the beam is used;
- using a polynomial of the first-order improves the localizing performance of the TRGS method;
- the procedure of MO improves the localizing performance of the TRGS method.

Acknowledgements

Sampaio, Silva, and Maia acknowledge the support of the Navy Research Center, CINAV, Portuguese Navy, the Portuguese Foundation for Science and Technology, FCT, under LAETA, through IDMEC, and the strategic project PEst-OE/EME/UI0667/2014 (UNIDEMI, FCT-NOVA). Zhong acknowledges the National Natural Science Foundation of China (51675103), the Fujian Provincial Excellent Young Scientist Fund (2014J07007) and the Specialized Research Fund for the Doctoral Program of Higher Education, the Ministry of Education, P. R. China (20133514110008).

References

1. Farrar, C. R. and Doebling, S. W. *An Overview of Modal-Based Damage Identification Methods.* Engineering Analysis Group Los Alamos National Laboratory Los Alamos, NM, USA (1997).
2. Doebling, S. W., Farrar, C. R., Prime, M. B., and Shevitz, D. W. *Damage Identification and Health Monitoring of Structural and Mechanical Systems from Changes in their Vibration Characteristics: A Literature Review.* Los Alamos National Laboratory, LA-13070-MS, USA (1996).

3. Ewins, D. J. *Modal Testing: Theory and Practice*. Research Studies Press Ltd. (1994).

4. Beer, F. P. and Johnston, E. R. *Mechanics of Materials*, 2nd edn. McGraw-Hill (1989).

5. Farrar, C. R. and Doebling, S. W. The state of the art in vibration-based structural damage identification, 2-day short course, Madrid, Spain (5–6 June 2000).

6. Maia, Silva, He, Lieven, Lin, Skingle, To and Urgueira, Theoretical and Experimental Modal Analysis, Research Studies Press Ltd. (1997).

7. Hart, G. C. and Wong, K. *Structural Dynamics for Structural Engineers*. John Wiley & Sons (1999).

8. Heylen, W. and Janter, T. Extensions of the modal assurance criterion. *Transactions of the ASME* **112**, pp. 468–472 (1990).

9. Pascual, R., Golinval, J.-C., and Razeto, M. A frequency domain correlation technique for model correlation and updating. In *Proceedings of the XV International Modal Analysis Conference*, pp. 587–592, Orlando, USA (1997).

10. Pascual, R., Golinval, J.-C., and Razeto, M. On-line damage assessment using operating deflection shapes. In *Proceedings of the XVII International Modal Analysis Conference*, pp. 238–243, Orlando, USA (1999).

11. Fotsch, D. and Ewins, D. J. Application of MAC in the frequency domain. In *Proceedings of the XVIII International Modal Analysis Conference*, pp. 1225–1231, Orlando, USA (2000).

12. Ewins, D. J. and Ho, Y. K. On the structural damage identification with mode shapes. In *Proceedings of the European COST F3 Conference on System Identification & Structural Health Monitoring*, pp. 677–684, Universidade Politécnica de Madrid, Spain (June 2000).

13. Pandey, A. K., Biswas, M., and Samman, M. M. Damage detection from changes in curvature mode shapes. *Journal of Sound and Vibration* **145**(2), pp. 321–332 (1991).

14. Stubbs, N., Kim, J. T., and Farrar, C. R. Field verification of a nondestructive damage localization and severity estimator algorithm. In *Proceedings of the XIII International Modal Analysis Conference*, pp. 210–218, Nashville, USA (1995).

15. Sampaio, R. P. C., Maia, N. M. M., and Silva, J. M. M. Damage detection using the frequency-response-function curvature method. *Journal of Sound and Vibration* **226**(5), pp. 1029–1042 (1999).

16. Petro, S. H., Chen, S., GangaRao, H. V. S., and Venkatappa, S. Damage detection using vibration measurements. In *Proceedings of the XV International Modal Analysis Conference*, pp. 113–119, Orlando, USA (1997).

17. Ratcliffe, C. P. Damage detection using a modified Laplacian operator on mode shape data. *Journal of Sound and Vibration* **204**(3), pp. 505–517 (1997).

18. Ratcliffe, C. P. A frequency and curvature based experimental method for locating damage in structures. *Journal of Vibration and Acoustics ASME* **122**(3), pp. 324–329 (2000).

19. Maia, N. M. M., Silva, J. M. M., Almas, E. A. M., and Sampaio, R. P. C. Damage detection in structures: From mode shape to frequency response

function methods. *Mechanical Systems and Signal Processing* **17**(3), pp. 489–498 (2003).

20. Maia, N. M. M., Silva, J. M. M., and Sampaio, R. P. C. Localization of damage using curvature of the frequency response functions. In *Proceedings of the XV International Modal Analysis Conference*, pp. 942–946, Orlando, USA (1997).

21. Sampaio, R. P. C., Maia, N. M. M., and Silva, J. M. M. The frequency domain assurance criterion as a tool for damage detection, Damage Assessment of Structures (DAMAS), pp. 69–76, Southampton, UK (2003).

22. Maia, N. M. M., Almeida, R. A. B., Urgueira, A. P. V., and Sampaio, R. P. C. Damage detection and quantification using transmissibility. *Mechanical Systems and Signal Processing* **25**, pp. 2475–2483 (2011).

23. Sampaio, R. P. C., Maia, N. M. M., Almeida, R. A. B., and Urgueira, A. P. V. A simple damage detection indicator using operational deflection shapes. *Mechanical Systems and Signal Processing* **72–73**, pp. 629–641 (2016).

24. Yoon, M. K., Heider, D., Gillespie Jr., J. W., Ratcliffe, C. P., and Crane, R. M. Local damage detection using the two- dimensional gapped smoothing method. *Journal of Sound and Vibration* **279**, pp. 119–139 (2005).

25. Dessi, D. and Camerlengo, G. Damage identification techniques via modal curvature analysis: Overview and comparison. *Mechanical Systems and Signal Processing* **52–53**, pp. 181–205 (2015).

26. Katz, M. J. Fractals and the analysis of waveforms. *Computers in Biology and Medicine* **18**(3), pp. 145–156 (1988).

27. Wahab, M. M. A. and Roeck, G. D. Damage detection in bridges using modal curvature: Application to a real damage scenario. *Journal of Sound and Vibration* **226**, pp. 217–235 (1999).

28. Heylen, W., Lammens, S., and Sas, P. *Modal Analysis Theory and Testing*, K. U. Leuven — PMA, Belgium, Section A.6. (1998).

29. Allemang, R. J. and Brown, D. L. A correlation coefficient for modal vector analysis. In *Proceedings of the 1st International Modal Analysis Conference*, pp. 110–116, Orlando, USA (1982).

30. Lieven, N. A. J. and Ewins, D. J. Spatial correlation of mode shapes, the coordinate modal assurance criterion (COMAC), In *Proceedings of the 6th International Modal Analysis Conference*, pp. 690–695, Kissimmee, USA (1988).

31. Ladevèze, P. Recalage de modélisations des structures complexes (Note technique 33.11.01.4). Tech. rep., Aérospatiale, Les Mureaux, France (1983).

32. Silva, T. A. N. and Maia, N. M. M. Detection and localisation of structural damage based on the error in the constitutive relations in dynamics. *Applied Mathematical Modelling* **46**, pp. 736–749 (2017).

Index

Computational and Experimental Methods in Structures

(Continued from page ii)

www.ingramcontent.com/pod-product-compliance
Lightning Source LLC
Chambersburg PA
CBHW050553190326
41458CB00007B/2031